THE SUGAR INDUSTRY
IN PERNAMBUCO

Modernization Without
Change, 1840-1910

Peter L. Eisenberg

University of California Press
Berkeley, Los Angeles, London

UNIVERSITY OF CALIFORNIA PRESS
BERKELEY AND LOS ANGELES, CALIFORNIA
UNIVERSITY OF CALIFORNIA PRESS, LTD.
LONDON, ENGLAND

COPYRIGHT © 1974, BY
THE REGENTS OF THE UNIVERSITY OF CALIFORNIA
ISBN: 0-520-01731-5
LIBRARY OF CONGRESS CATALOG CARD NUMBER: 75-117340
PRINTED IN THE UNITED STATES OF AMERICA

TO ROSA

CONTENTS

ILLUSTRATIONS

Maps

Pernambuco, 1910
Brazil, 1911

Plates

TABLES

PREFACE

*I*n this book I deal primarily with economic modernization and change. This phenomenon is usually defined as "the development of industrial systems based on high level of technology, on growing specialization of economic roles, and of units of economic activity—production, consumption, and marketing—and on the growth of the scope and complexity of the major markets, the markets for goods, labor, and money."[1] I contend that modernization in the sense of technological improvement and the reorganization of production, and the conversion from slave to free labor did not restore the health of the Pernambuco sugar industry. Moreover the agents of modernization, the sugar planters aided by the governments, used the process to shore up their own position in the local economy and society. Thus change, in terms of a new distribution of power and income, did not occur.

I originally conceived the study as a biography of a single mill, from traditional mill *(engenho)* to modern factory *(usina)*, but lack of data defeated this plan. I next attempted to replicate the impressive work of Stanley Stein on the coffee county of Vassouras with a study of a Pernambuco sugar county, Escada. Again a scarcity of documents frustrated my intention. I finally decided to focus on the sugar industry as a whole within the political unit of Pernambuco.

I begin the study in the decade of the 1840's. During this decade the British relaxed tariff restrictions and began importing Brazilian crude sugar in significant quantities. This British

1. S. N. Eisenstadt, *Modernization: Protest and Change* (Englewood Cliffs, N.J., 1966), pp. 3-4.

market would stimulate Brazilian production for several decades, until the entrance of European beet sugars in overseas
markets initiated a prolonged export crisis from which the industry never recovered. This decade also witnessed the last
major imports of African slaves to work the plantations. After
1850 the international slave trade came to a quick end, with
important repercussions in the domestic labor market.

I end the study in the first decade of the twentieth century.
By then the principal effects of the export crisis had been felt,
and the mill owners had made their most ambitious effort toward technological modernization. Moreover, slavery had been
abolished twenty years previously, and the modes of free labor
employment had become clearly established.

Part One of this book reviews the economic crisis of the later
nineteenth century and the planters' responses. It analyzes the
export trade, Pernambuco's inability to preserve major foreign
markets, and the failure of attempts to develop the Brazilian market. It then describes traditional sugar manufacture and the pace
of technological change. Part I concludes with two chapters
describing how some planters, aided by official subsidies, managed to transform their antiquated engenhos into modern usinas.

Part Two reviews the social crisis and the transition from
slave to free labor. We measure the planters' control of land and
its political consequences. We then describe how this power
enabled them not only to endure but also to benefit from the process of gradual abolition, and how they were able to convert to free
labor without paying higher wages and without subsidizing European immigration. Thus, the planters in large part compensated
for their difficulties in the market by sharing much of the cost of
stagnation with the workers. In my conclusion I reformulate the
main points of the analysis, try to place Pernambuco in perspective with other cane-sugar and slavery economies, and the rest of
Brazil, and consider whether any alternate historical patterns
could have emerged.

Most histories follow a chronological progression, beginning
with the more remote events and tracing developments to a more

recent period. Each of our chapters describes such a progression, from the middle of the nineteenth century to the beginning of the twentieth. But each succeeding chapter begins again in the middle nineteenth century. Each chapter thus dissects a different aspect of the sugar economy over the same period. We tried and deliberately rejected a strictly chronological approach because the dominant stability during the period would have burdened such an organization with constant repetition.

This book should contribute to several branches of Brazilian history. The Brazilian sugar history, especially that of the principal producing regions, has been treated by Miguel Costa Filho, Gileno Dé Carli, Manuel Diégues Júnior, J. H. Galloway, Moacyr Medeiros de Sant'Ana, Alberto Ribeiro Lamego, Maria Teresa Schorer Petrone, and José Araujo de Wanderley Pinho; our study clarifies the nature of transitions in later nineteenth-century Pernambuco. The history of slaves in the sugar industry and the Brazilian abolition of slavery in general has also attracted numerous Brazilian and foreign scholars, including Fernando de Azevedo, Leslie Bethell, Robert Conrad, Gilberto Freyre, Mauricio Goulart, Evaristo de Moraes, Roberto Simonsen, Robert Brent Toplin, and Emilia Viotti da Costa.[2] More than most, our study emphasizes the benefits derived by the planters and the costs incurred by the ex-slaves and other free workers as a result of abolition.

In a broader perspective, our study adds to the literature on the development and underdevelopment of tropical plantation societies in the Western Hemisphere. Although we did not make explicit comparisons, there are obvious parallels between Brazil's cane sugar economy and society and those, for example, of Cuba,

2. Miguel Costa Filho, *A Cana-de-Açúcar em Minas Gerais* (Rio de Janeiro, 1963). Manuel Diégues Júnior, *O Bangué nas Alagôas* (Rio de Janerio, 1949). Moacyr Medeiros de Sant'Ana, *Contribuição à História do Açúcar em Alagôas* (Recife, 1970). Alberto Ribeiro Lamego, *Homem e o Brejo* (Rio de Janeiro, 1945). Maria Teresa Schorer Petrone, *A lavoura canavieira em São Paulo* (São Paulo, 1968). For works by Dé Carlí, Galloway, Wanderley Pinho, and the slavery historians, see the Bibliography. Dé Carlí has also written *Evolução do problema canavieiro fluminense* (rio de Janeiro, 1942) and *Gênese e evolução da indústria açucareira paulista* (Rio de Janeiro, 1943).

Louisiana, Puerto Rico, and Jamaica. The factors conditioning the export cycles and the impacts of technological and social change were often similar, and our conclusions about Pernambuco undoubtedly are relevant elsewhere.

This study has its own history, and I owe its completion to the help of many persons. My interest in Pernambuco and sugar was first aroused when I read Gilberto Freyre's *The Masters and the Slaves*. In 1965 Charles Wagley suggested the transition from engenho to usina as a doctoral dissertation topic. Lewis Hanke watched the evolution of the dissertation in its earliest stages, and E. Bradford Burns guided the writing and defense in 1969. I am grateful to both. At the defense Stuart Bruchey, Nathaniel Leff, Dwight Miner, and Ronald Schneider made valuable criticisms. If the book differs from the dissertation, many of the changes for the better can be attributed to David Denslow, Albert Fishlow, Herbert Klein, Stuart Schwartz, Thomas Skidmore, and Stanley Stein. My colleagues at Rutgers, especially Michael Adas, Samuel Baily, Tilden Edelstein, Karl Hardach, and David Ringrose, commented helpfully on certain chapters.

Like other North American researchers, I have enjoyed and profited from the disinterested good will of Brazilian archival and library personnel. I would particularly like to thank the following persons for their help: Maria Walda de Aragão Araújo and Pedro Moniz de Aragão of the Arquivo Nacional; José Bezerra of the Assembléa Legislativa de Pernambuco; Olimpio Costa Júnior of the Pernambuco Biblioteca Pública do Estado; Jordão Emerenciano and Lúcia Nery da Fonesca of the Pernambuco Arquivo Público Estadual; Walfrido Cezar Freire of the Rede Ferroviária do Nordeste; Antônio Galvão of the Associação Commercial de Pernambuco; José Hipólito de Monteiro of the Cartório Público de Ipojuca; Luis Oiticica of the Museu do Açúcar; Xavier Placer of the Ministério de Agricultura; and Nair Rodrigues of the Departmento de Obras Públicas de Pernambuco. Manuel Cardozo of the Oliveira Lima Library in Washington, D.C., and Samuel Schoenfeld of Lamborn and Company, Inc., in New

York, also greatly facilitated my work in their respective institutions.

In Recife I received the cooperation of Ayrton and Rui Cardoso, Antiógenes Chaves, Luis Dias Lins, Ilvo Meirelles, Ricardo Pessôa de Queiroz, and Cid Sampaio, all usineiros with a sense of history, as well as Samuel da Silva Costa and Fernando Perez Garcia of Mendes Lima Cia. I am also indebted, intellectually and otherwise, to Antônio Falcão de Albuquerque Maranhão, Alexandre José Barbosa Lima Sobrinho, Amaro Cavalcanti, Paulo Cavalcanti, Robert Conrad, Manuel Correia de Andrade, Miguel Costa Filho, Roger Cuniff, Gileno Dé Carli, Manuel Diégues Júnior, John Dumoulin, Ludlow Flower, Gilberto Freyre, José Antônio Gonsalves de Mello Neto, Richard Graham, Louisa Schell Hoeberman, Robert Kestell, John Knodel, Betsy Kuznesof, Nício de Lima Barbosa and Claribalte Passos of *Brasil Açucareiro*, Gil Maranhão, Maria Laura Menezes, Mirocem Navarro, Amaro Quintas, Jaime Reis, Paul Silberstein, Ênio Silveira, Robert Slenes, Pelópidas Soares, and Nelson Werneck Sodré. Flávio Guerra, Luis de Nascimento, Maria Tereza Schorer Petrone, and Paul Singer were kind enough to let me read their manuscripts before publication.

This book represents a return on an investment of many thousands of dollars. A large part of this investment was granted by the Office of Education, U.S.. Department of Health, Education, and Welfare; Resources for the Future, Inc.; and the Research Council of Rutgers University. I appreciate the confidence of these institutions; of course they cannot bear any responsibility for my statements, opinions, or conclusions.

I wish also to thank my parents, Monroe and Winona Eisenberg, whose support and encouragement maintained me during seemingly endless years of graduate studies. In Olinda and Niterói I enjoyed the generous hospitality of my wife's family; Luis and Nicette Almeida, Ywalter and Zélia Gusmão, Hortêncio and Socorro Navarro de Mesquita; and Geraldo and Marta Reis.

I dedicate the book to my wife, Maria do Rosário Navarro de

Oliveira Eisenberg, whom I met in the course of research and without whom perhaps neither the writing nor the publication would have come to pass. If such a study is a labor of love, one could only expect a wife to become jealous. Rosa worked happily with me for several years despite the sacrifices; her collaboration has been essential.

The Sugar Industry in Pernambuco

ABBREVIATIONS

AP	Accounts and Papers
APE	Arquivo Público Estadual
ACA	Associação Commercial Agricola
ACBP	Associação Comercial Beneficente de Pernambuco
CSFB	Central Sugar Factories of Brazil, Limited
HCC	House of Commons by Command
IIPA	Imperial Instituto Pernambucano de Agricultura
IAA	Instituto do Açúcar e do Alcool
NBSF	North Brazilian Sugar Factories, Limited
SAAP	Sociedade Auxiliadora da Agricultura de Pernambuco
SNA	Sociedade Nacional da Agricultura
USAP	União dos Syndicatos Agricolas de Pernambuco

Part One

THE ECONOMIC CRISIS

Antes foram engenhos,
poucos agora são usinas.
Antes foram engenhos,
agora são imensos partidos.
Antes foram engenhos
com suas caldeiras vivas;
agora são informes
corpos que nada identifica.

"O Rio," João Cabral de Melo
Neto, in *Morte e Vida Severina
e Outros Poemas em Voz Alta*
(Rio de Janeiro, 1966), p. 135.

INTRODUCTION:
THE COLONIAL HERITAGE OF
NINETEENTH-CENTURY BRAZIL

*A*lthough Brazil gained political independence from Portugal in 1822, the nineteenth-century economic experience of this largest Latin American nation was in many ways an extension of its colonial past. Even the more radical breaks in century-old patterns, such as the opening of the ports to trade with all nations and the abolition of slavery, did not soon alter the nature of basic relationships between Brazil and northern hemisphere countries, or between the owners of the means of production and the labor force.[1]

The Colonial Heritage

From the time of its discovery in 1500, when the Portuguese first established trading posts for dyewood *(pau brasil)* on the endless beaches between the present-day states of Rio de Janeiro and Rio Grande do Norte, Brazil has exported raw materials to European markets. The succession of Brazilian export cycles has been described to English language readers often enough in recent years to make repetition unnecessary here.[2] We need only note that the sugar cycle, a later phase of which is the subject of this

---

1. Stanley J. Stein and Barbara H. Stein, *The Colonial Heritage of Latin America, Essays on Economic Dependence in Perspective* (New York, 1970), pp. 147-150, present a recent formulation of this thesis.
2. Celso Furtado, *The Economic Growth of Brazil, A Survey from Colonial to Modern Times*, translated by Ricardo W. de Aguiar and Eric Charles Drysdale (Berkeley and Los Angeles, 1963, first published as *Formação Econômica do Brasil*, 1959); Rollie Poppino, *Brazil, The Land and the People* (New York, 1968); E. Bradford Burns, *A History of Brazil* (New York, 1971).

book, began shortly before 1550, and ended, according to most economic historians, in the mid-seventeenth century when Dutch, British, and French West Indian colonies in the Caribbean, by virtue of their proximity to Europe, colonial preference, and natural advantages, squeezed Brazilian sugars out of traditional markets.

But one should not forget two aspects of this cycle, which were common also to most of the other export cycles. First, the sugar cycle did not end abruptly in 1650. Although exports declined for the next 150 years, by the early nineteenth century sugar exports were again increasing, and they continued to grow until the turn of the twentieth century. Secondly, the Brazilian export cycle can be defined as the period when a particular product dominated both the country's export list and the world market in that commodity, with the results that it attracted the movable productive factors—that is, capital and labor—and caused a new distribution of income both vertically, among classes, and horizontally, among geographical regions.[3] But quite frequently the export activity reached a much higher level—both in terms of quantity and value of product and in terms of amounts of capital and labor employed—at a much later date. Thus despite declining sugar export volumes during the eighteenth century, the value earned in that activity almost continually exceeded the value earned in gold and diamond mining during the Golden Age up to 1760.[4] Similarly, during the nineteenth century, although coffee outranked sugar after 1830, the absolute volume of sugar exports reached a level 500 percent above the apogee of the colonial sugar cycle (Table 1).[5]

3. Mircea Buescu and Vicente Tapajós, *História do Desenvolvimento Econômico do Brasil* (Rio de Janeiro, n.d.), pp. 24-25, include some elements of this definition.

4. C.R. Boxer, *The Golden Age of Brazil, 1695-1750. Growing Pains of a Colonial Society* (Berkeley, 1961). Brazilian economic historians have pointed out that the mining cycle only reached its peak around 1760, contrary to the impression created by Boxer. Furtado, *The Economic Growth of Brazil*, p. 85. Buescu and Tapajós, *História*, p. 42.

5. Roberto C. Simonsen, *História Econômica do Brasil, 1500-1820*, 4th edition (São Paulo, 1962, first published 1937), table facing p. 382. "O Açucar na vida econômica do Brasil," *Annuario Açucareiro para 1938*, (Rio de Janeiro), p. 235.

TABLE 1
VALUE OF SELECTED BRAZILIAN EXPORT
PRODUCTS IN RELATION TO TOTAL EXPORTS
(in percent)

Years	Sugar	Gold and Diamonds	Cotton	Coffee	Hides	To-bacco	Cocoa	Natural Rubber	Mate Tea
1500-1822	56.0	31.7	2.2	0.7	2.8	2.2	0.7		
1821-30	30.1		20.6	18.4	13.6	2.5	0.5	0.1	
1831-40	24.0		10.8	43.8	7.9	1.9	0.6	0.3	0.5
1841-50	26.7		7.5	41.4	8.5	1.8	1.0	0.4	0.9
1851-60	21.2		7.5	48.8	7.2	2.6	1.0	2.3	1.6
1861-70	12.3		18.3	45.4	6.0	3.0	0.9	3.1	1.2
1871-80	11.8		9.5	56.6	5.6	3.4	1.2	5.5	1.5
1881-90	9.9		4.2	61.5	3.2			8.0	
1891-1900	6.0		2.4	64.5			1.5	15.8	
1901-1910	1.2		2.1	51.3			2.8	27.9	

SOURCES: Simonsen, *História Econômica do Brasil*, p. 381. Nelson Werneck Sodré, *História da Burguesia Brasileira*, 2nd ed. (Rio de Janeiro, 1967) pp. 62-104. Buescu and Tapajós, *História do Desenvolvimento Econômico do Brasil*, p. 28. Virgílio Noya Pinto, "Balanço das Transformações Econômicas no Século XIX," in Carlos Guilherme Mota (ed.), *Brasil em Perspectiva* (São Paulo, 1968), p. 139.

Three other aspects of the colonial heritage deserve mention for their continued importance after independence: monopsony/monopoly, latifundia, and slavery. During most of the colonial period, exporters were obliged to sell in Portuguese markets. The metropolitan authorities imposed this obligation in order to guarantee tax collections on colonial products, and to favor home merchants, who then resold Brazilian goods in western European markets.[6] The same mercantilist commercial policy obliged Brazilians to buy all their imports from Portugal, which acted for most of the colonial period as an entrepôt for northern European manufacturers—particularly the English, whose Treaty of Methuen (1703) granted tariff reductions on Portuguese wine in exchange for similar favors on English textiles.[7] For Brazil, this policy had clear disadvantages: the exporters did not receive the higher prices paid in the consumer markets, the importers had to pay transshipment costs through Portugal, and the entrepreneurs were prevented from establishing local industries by the metropolitan commercial interests.

Many agricultural export commodities were most profitably grown on large plantations. Cotton, coffee, and especially sugar required large-scale production units: the small family-operated farm could not keep unit costs below world prices. Moreover, land distribution in the colony had traditionally depended upon royal grants of huge parcels of one square league or more. The Portuguese monarchs had distributed these *sesmarias* (land grants) to preferred colonists who would occupy and defend the land against foreign intruders, and who also would develop the land as a source of taxable exports. These large grants, augmented by additional concessions, inheritances, and purchases, dominated the principal areas of agricultural exports; small properties prevailed only in the peripheral areas of livestock, subsistence crops, and cer-

6. Helio Jaguaribe, *Economic and Political Development. A Theoretical Approach and a Brazilian Case Study* (Cambridge, Mass., 1968, first published as *Desenvolvimento Econômico e Desenvolvimento Político*, 1962), pp. 107-112.

7. Furtado, *The Economic Growth of Brazil*, pp. 88-89. See Fernando A. Novais, "O Brasil nos Quadros do Antigo Sistema Colonial," in Carlos Guilherme Mota, *Brasil em Perspectiva*. (São Paulo, 1968), pp. 51ff., for a synthesis of Portuguese mercantilism with bibliography.

tain exports such as tobacco.[8] This land tenure system impeded
the development of domestic food production, concentrated the
colony's income in the small group of large landowners and the
commercial community, and subjugated the rural populations to
the hegemony of the landed elite.

The reliance on slave labor, predominantly African in origin,
came about because of the absence in Brazil of large populations
of Amerindians engaged in sedentary agriculture, and because of
the inability of the European immigrants, whether by tempera-
ment or by choice, to satisfy the demand for agricultural labor.
The importation of millions of Africans in bondage met that
demand, but it also had two serious disadvantages for free Brazilians.[9] Slaves were not active consumers, so the domestic market
remained quite small; and the slaves' presence depressed wages
and inhibited the development of a free rural proletariat in most
regions.

Nineteenth-Century Brazil

In the first years of independence, most of the products of
earlier export cycles still dominated the new nation's foreign
trade. Sugar, produced in the northeastern provinces and particularly in Pernambuco and Bahia, led the lists after Haiti's independence revolutions (1801-05) and Napoleon's continental blockade (1805-14) largely deprived European consumers of Caribbean sugars and raised sugar prices. Cotton, encouraged in the later
eighteenth century by the Marques de Pombal's monopoly company in the northern provinces of Pará and Maranhão, benefited
from the United States' war of independence (1775-83), the
Embargo and Non-Intercourse Acts (1807-10) and the War of
1812 (1812-15), which deprived English textile mills of North
American cotton and allowed Brazilian cotton to reach second
place on the export list. The third major export, coffee, was the

8. Caio Prado Júnior, *Formação do Brasil Contemporâneo: Colônia*, 5th
edition (São Paulo, 1957, first published 1942), pp. 147-149.
9. The number of Africans imported is subject to debate. Simonsen estimated 3.5 million, a figure accepted by many other scholars. Mircea Buescu,
História Econômica do Brasil, Pesquisas e Análises (Rio de Janeiro, 1970), pp.
201-218, discusses these estimates, and makes his own calculation of 5 to 5.5 million.

only newcomer to the list. Like sugar, coffee profited from the withdrawal of Haitian supply from world markets; the principal coffee export plantations were located in Rio de Janeiro province.

It was the coffee boom that caused the principal economic changes in nineteenth century Brazil. After 1830, coffee earned more foreign exchange than any other export, and its lead grew more or less steadily from then on (Table 2). Rio de Janeiro, São Paulo, and Minas Gerais, the principal coffee-growing provinces in the center-south, drained slaves from the northeast after 1850, when the international slave traffic stopped, and attracted immigrants and capital from Europe after 1880. These movements increased the concentration of income and population in the center-south (relative to the northeast), created a mass market, and permitted the industrialization process to begin.[10]

Brazil's political independence from Portugal did not bring with it rapid diversification of foreign markets. When the Portuguese court fled Napoleon's armies to Brazil, King João VI upon arriving in the colony hastened to open the ports to the commerce of all nations. But rather than being a repudiation of colonial mercantilist policy, his act represented only the recognition that the colony could not trade with Portugal as long as a hostile power occupied the metropolis. In 1810, in payment for English protection from Napoleon, João VI signed a trade treaty giving the English preferential tariffs and extraterritorial rights in Brazil. Thus for strategic reasons Portugal preserved Brazil's dependence upon a single European market, the same market that had supplied the colony, through the Portuguese middleman, since 1703.

João's son Pedro I respected this treaty even when he declared independence from Portugal in 1822. When the trade treaty expired in 1825, Pedro I was obliged to sign a similar agreement with England to obtain the diplomatic recognition necessary for credits and trade. Both treaties tied Brazil's com-

10. Warren Dean, *The Industrialization of São Paulo, 1880-1945* (Austin, Texas, 1969), pp. 83ff. Nathaniel H. Leff, "Desenvolvimento econômico e desigualdade regional: Origens do caso brasileiro," *Revista brasileira de economia* (Rio de Janeiro), ano 26, no. 1 (January 1972), pp. 3-21. This shift in income and population away from the northeast had already begun during the eighteenth century mining boom, which attracted capital and slaves from the northeast to Minas Gerais, and caused the transferal of the colonial capital from Salvador to Rio de Janeiro.

TABLE 2
BRAZILIAN SUGAR AND COFFEE EXPORTS, (1821-1910)

Years	Sugar				Coffee			
	Annual Tons	Average £ Value [a]	Percent Total Export Value		Annual Tons	Average £ Value [a]	Percent Total Export Value	
1821-25	41,174	983,600	23.2		12,480	739,600	17.6	
1826-30	54,796	1,369,600	37.8		25,680	698,200	19.7	
1831-35	66,716	1,091,500	23.5		46,980	2,001,500	40.7	
1836-40	79,010	1,320,800	24.3		69,900	2,428,000	46.0	
1841-45	87,979	1,264,600	21.6		85,320	2,058,200	42.0	
1846-50	112,830	1,650,600	27.5		120,120	2,472,800	40.9	
1851-55	127,874	1,882,200	21.5		150,840	4,113,000	48.6	
1856-60	98,864	2,445,400	21.2		164,160	5,635,000	48.7	
1861-65	113,551	1,943,600	14.0		153,300	6,863,400	49.3	
1866-70	109,001	1,717,800	10.7		192,840	6,737,400	42.5	
1871-75	169,337	2,353,400	11.8		216,120	10,487,800	52.0	
1876-80	167,761	2,354,600	11.8		219,900	12,103,000	60.7	

(Continued)

Years	Sugar			Coffee		
	Annual Tons	Average £Value[a]	Percent Total Export Value	Annual Tons	Average £Value[a]	Percent Total Export Value
1881-85	238,074	2,646,000	13.7	311,760	11,359,000	58.8
1886-90	147,274	1,537,200	7.0	307,800	14,380,800	64.5
1891-95	153,333	2,182,800	7.2	361,092	20,914,000	69.2
1896-1900	113,908	1,288,800	4.7	532,800	16,669,400	60.4
1901-05	78,284	637,000	1.6	740,280	20,952,200	53.0
1906-10	51,338	479,600	0.8	826,908	27,877,000	50.5

[a] Nominal value unadjusted for inflation.

SOURCES: "O Açucar na vida econômica do Brasil," pp. 233-236. Affonso de Taunay, Pequena História do Café no Brasil (1727-1937) (Rio de Janeiro, 1945), pp. 547-549.

TABLE 2
BRAZILIAN SUGAR AND COFFEE EXPORTS, (1821-1910)

Years	Sugar			Coffee		
	Annual Tons	Average £ Value [a]	Percent Total Export Value	Annual Tons	Average £ Value [a]	Percent Total Export Value
1821-25	41,174	983,600	23.2	12,480	739,600	17.6
1826-30	54,796	1,369,600	37.8	25,680	698,200	19.7
1831-35	66,716	1,091,500	23.5	46,980	2,001,500	40.7
1836-40	79,010	1,320,800	24.3	69,900	2,428,000	46.0
1841-45	87,979	1,264,600	21.6	85,320	2,058,200	42.0
1846-50	112,830	1,650,600	27.5	120,120	2,472,800	40.9
1851-55	127,874	1,882,200	21.5	150,840	4,113,000	48.6
1856-60	98,864	2,445,400	21.2	164,160	5,635,000	48.7
1861-65	113,551	1,943,600	14.0	153,300	6,863,400	49.3
1866-70	109,001	1,717,800	10.7	192,840	6,737,400	42.5
1871-75	169,337	2,353,400	11.8	216,120	10,487,800	52.0
1876-80	167,761	2,354,600	11.8	219,900	12,103,000	60.7

(Continued)

Years	Sugar			Coffee		
	Annual Tons	Average £Value[a]	Percent Total Export Value	Annual Tons	Average £Value[a]	Percent Total Export Value
1881-85	238,074	2,646,000	13.7	311,760	11,359,000	58.8
1886-90	147,274	1,537,200	7.0	307,800	14,380,800	64.5
1891-95	153,333	2,182,800	7.2	361,092	20,914,000	69.2
1896-1900	113,908	1,288,800	4.7	532,800	16,669,400	60.4
1901-05	78,284	637,000	1.6	740,280	20,952,200	53.0
1906-10	51,338	479,600	0.8	826,908	27,877,000	50.5

[a] Nominal value unadjusted for inflation.

SOURCES: "O Açucar na vida econômica do Brasil," pp. 233-236. Affonso de Taunay, *Pequena História do Café no Brasil (1727-1937)* (Rio de Janeiro, 1945), pp. 547-549.

markets. In the center-south, where the ex-slaves might have aspired to a share of the coffee prosperity, the arrival of thousands of Europeans immigrants defeated that hope.

This rather dismal summary of continuities between colonial and independent Brazil should not leave the impression of a completely static history. Especially during the Second Empire (1840-89), the growth of large urban areas such as Rio de Janeiro and São Paulo, the adoption of technological innovations such as the steam engine and the railroad, the appearence of consumer goods industries, and even an emerging entrepreneurial mentality, marked the beginning of a new industrial era.[15] The northeastern sugar areas succeeded, with government aid, in modernizing mill technology and increasing production scale, and to this extent at least they matched the progress being made in the center-south. But the northeastern development pales quickly in comparison with the coffee-growing areas; in effect, the sugar-producing areas stagnated.

15. Dean, *The Industrialization of São Paulo,* chapter 1. Richard Graham, *Britain and the Onset of Modernization in Brazil, 1850-1914* (Cambridge, Eng., 1968), pp. 9-22.

2

ECONOMIC CRISIS: THE DECLINE OF EXPORTS

$\mathscr{D}$uring the later nineteenth century, the Pernambuco sugar industry led Brazil in exports and exemplified well the problems of the national industry. Along with most other cane sugar producers of the world, Pernambuco experienced two kinds of difficulties: falling prices and stiffer competition. The Brazilians failed to overcome these difficulties and their industry stagnated. Deteriorating export revenues first indicated the industry's predicament. Falling sugar prices, especially after 1860, reduced returns in the early 1870's to the level of the 1850's (Tables 3 and 4). Total production volume increased fairly steadily during the century in response to European and North American population growth and short-run price elastic demand—that is, the producers could not raise prices by withholding supply so they made as much sugar as possible for a given price, and as a result of sugar's continuing comparative advantage in Pernambuco, which attracted new entries by yielding profits higher than those gained by any other use of land in the province.[1] In the 1880's access to the U.S. market spurred export growth, but by the 1890's sugar exports were again in serious trouble (Table 5).

1. Only rarely did the sugar prices fall so low that planters preferred to abandon or burn their cane rather than incur manufacturing costs. Report by Hughes, Pernambuco, April 30, 1885, in Great Britain, Parliament, *Parliamentary Papers* (House of Commons by Command), 1884-85, v. LXXIX, *Accounts and Papers*, v. XVIII, p. 1,323. Later references to these Papers will be abbreviated HCC, volume number, *AP*, volume number.

TABLE 3
PERNAMBUCO SUGAR EXPORTS

Years [a]	Average Annual Quantity (tons)	Average Annual Value (£1880) [b]
1836-40	27,844	306,881
1841-45	31,926	409,708
1846-50	47,932	634,628
1851-55	56,981	949,453
1856-60	48,523	1,007,331
1861-65	46,741	698,008
1866-70	63,229	748,455
1871-75	78,699	930,345
1876-80	91,882	1,280,670
1881-85	103,889	1,188,376
1886-90	119,227	1,590,118
1891-95	n.a.	n.a.
1896-1900	40,840	284,079
1901-05	11,701	141,299
1906-10	32,993	361,517

[a] Unless otherwise noted, all harvest years have been assumed identical to the first calendar year, for example, 1836-37 was calculated as 1836.

[b] Real value was calculated by converting mil-réis to nominal pounds at exchange rates given in Ónody, and then dividing nominal pounds by a British export price index in terms of 1880 prices given in Imlah.

SOURCES: George Eduardo Fairbanks, *Observações sôbre o Commércio de Assucar, e o Estado presente desta Indústria* (Bahia, 1847), p. 149. Sebastião Ferreira Soares, *Notas Estatísticas sôbre a Produção Agrícola e Carestia dos Gêneros Alimentícios no Império do Brasil* (Rio de Janeiro, 1860), pp. 255-256. *Informações sôbre o Estado de Lavoura* (Rio de Janeiro, 1874), p. 162. *Relatório da Direcçao da Associaçao Commercial Beneficente de Pernambuco apresentado à Assembléa Geral da Mesma em 25 de agôsto de 1862*, table. *Falla com que o Exmo. Sr. Doutor Manoel Clementino Carneiro da Cunha abrio a sessão da Assembléa Legislativa Provincial de Pernambuco em 2 de março de 1877*, "Registro da Alfandega." *Relatório da Direcção da Associaçao Commercial Beneficente de Pernambuco apresentado à Assembléa Geral da mesma em 8 de agôsto de 1884*, table. *Relatório da Direcção da Associação Commercial Beneficente de Pernambuco apresentado à Assembléa Geral da mesma em 9 de agôsto de 1889*, table. João Severiano da Fonseca Hermes Júnior, *O Assucar como factor importante da riqueza pública no Brasil* (Rio de Janeiro, 1922), pp. 49-50, 71-72. *Economia e Agricultura*, ano 1, no. 5 (February 15, 1933), p. 38. Albert H. Imlah, *Economic Elements in the Pax Britannica, Studies in British Foreign Trade in the Nineteenth Century* (Cambridge, Mass., 1958), Table 8. Oliver Ónody, *A Inflação Brasileira (1820-1958)* (Rio de Janeiro, 1960), pp. 22-23.

TABLE 4

SUGAR VALUES AND FOREIGN EXCHANGE RATES

Years	Mascavado Price (Shillings/cwt. c.i.f. London)	Exports (£/ton Brazil)	Exchange Rate (d/1$000, Rio de Janeiro)
1836-40	40/5	15.7	31.7
1841-45	36/0	14.5	26.7
1846-50	26/2	14.3	26.9
1851-55	21/10	15.9	28.1
1856-60	26/7	20.9	26.1
1861-65	22/2	17.1	26.2
1866-70	22/5	15.4	20.9
1871-75	23/0	13.7	25.6
1876-80	21/1	14.4	23.3
1881-85	17/5	10.5	20.8
1886-90	13/1	11.3	23.1
1891-95	12/6	14.5	11.7
1896-1900	10/4	11.8	8.3
1901-05	9/4	9.4	12.7
1906-10	9/10	9.7	15.6

NOTE: The mascavado price decline is more violent than the decline in average value per ton exported because the latter included other higher-priced qualities of sugar.

SOURCES: Noel Deerr, *The History of Sugar*, 2 vols. (London, 1949-50), v. II, p. 531. "O Açucar na vida econômica do Brasil," pp. 233-235. Ónody, *A Inflação Brasileira*, pp. 22-23.

TABLE 5

TOTAL PERNAMBUCO SUGAR PRODUCTION

Years	Average Annual Quantity (tons)	Years	Average Annual Quantity (tons)
1801-05	8,362	1856-60	67,339
1806-10	7,866	1861-65	57,357
1811-15	5,742	1866-70	54,372
1816-20	9,198	1871-75	98,231
1821-25	12,212	1876-80	116,379
1826-30	18,234	1881-85	133,847
1831-35	13,690	1886-90	156,321
1836-40	26,743	1891-95	173,442

Years	Average Annual Quantity (tons)	Years	Average Annual Quantity (tons)
1841-45	32,357	1896-1900	134,326
1846-50	49,925	1901-05	142,015
1851-55	63,312	1906-10	141,624

NOTE: Because of discrepancies between this table and table 3—for example, exports for the periods 1836-40 and 1866-70 exceeded sugar entries to Recife—this table serves only to suggest relative changes in production.

SOURCE: Gaspar Peres and Apollonio Peres, *A Industria Assucareira em Pernambuco* (Recife, 1915), pp. 27-29, 31, 109-110, 114. The authors used Associação Commercial Beneficente de Pernambuco records of sugar entering Recife. These figures would not necessarily correspond to actual production, for they omitted sugar produced in Pernambuco mills near the northern and southern borders and marketed through João Pessôa or Maceió, respectively, and included sugar produced elsewhere and marketed through Recife.

Exports

Negative trends in export trade hurt the industry because foreign markets absorbed over three-quarters of Pernambuco's sugar (Table 6). Moreover, the consequences of such adverse turns were magnified by Brazil's high dependence upon imported goods, which required the foreign exchange earned by selling

TABLE 6

PERNAMBUCO'S AVERAGE ANNUAL
SUGAR EXPORTS AS PERCENTAGE OF
AVERAGE ANNUAL PRODUCTION

Years	Export quantity/ total quantity (percent)	Export value/ total value (percent)
1856-60	79.9	76.2
1861-65	81.5	77.5
1866-70	85.8	81.8
1871-75	85.5	82.5
1876-80	84.2	79.8
1881-85	82.8	78.0
1886-90	81.4	77.5
1897-1901	45.9	
1903-05	9.1	8.9
1906-10	24.8	15.2

SOURCES: Tables 3, 5 and 9.

abroad. The falling foreign exchange rate after 1851 only compounded the problem. Par had been set at 27d/1$000 in 1846, but the free market rate rarely reached this level. Its decline in the second half of the century accelerated under the monetary policies of the First Republic and reached an unprecedented low of 7.4d in the later 1890's (Table 4).

This decline in the foreign exchange rate had two contrasting effects. On the one hand, it benefited the sugar producers because it enabled them to sell foreign currency earned in exports for increasing amounts of mil-réis. Thus declines in the exchange rate could compensate for drops in sugar prices. Spokesmen for Pernambuco sugar interests criticized the Imperial policy which established a high par as "a death sentence for the province's cane growing and sugar industry," and urged the government to refrain from foreign loans which raised the exchange rate.[2]

But if a falling exchange rate helped exporters, it hurt importers, for whom foreign exchange became more expensive. The sugar producers were both exporters and importers, since nearly all their capital equipment, as well as many of their consumption goods, came from abroad. Thus the net effect of a falling exchange rate depended in large part upon the planter's ratio of imports to exports, and on the degree to which domestically produced goods reflected import price increases. Unfortunately for the planters, the modernization of sugar technology in the later nineteenth century required large investments in imported machinery to increase production scale and efficiency. For those producers who tried to remain competitive by modernizing, therefore, the falling exchange rate was a handicap. During the 1880's, the period of most rapidly falling exchange rates, the state government had to subsidize the modernization which private individuals were reluctant to undertake. Moreover, even those

2. Sociedade Auxiliadora da Agricultura de Pernambuco, Livro de Atas no. 2, Conselho Administrativo, October 27, 1886. *Idem.*, "Relatório da Sessão de 28 de setembro," *Diário de Pernambuco* (Recife), October 1, 1886. "Crise do Assucar," *ibid.*, August 19, 1887. Henrique Augusto Milet, *Auxílio à Lavoura e Crédito Real* (Recife, 1876), pp. 73-74. The imperial government established a high par exchange rate, and contracted foreign loans, in order to buy gold more cheaply to repay older obligations.

planters who did not modernize suffered from exchange movements, for while sugar prices dropped to less than one-fourth, the exchange rate declined only by one-half. Thus the exchange rate falls did not compensate for the sugar price falls.

The growing European beet sugar industry caused many of Pernambuco's problems. That industry, born in the early nineteenth century to compensate for the exclusion of colonial cane sugar by Napoleon's continental blockade, gained government protection against cane sugar imports when the blockade ended, and it grew to satisfy domestic needs. The beet sugar producers soon sought foreign markets, and the world price fell rapidly.[3] By 1900, producers of crude sugar were earning less than one-fourth the price paid sixty years earlier. The beet sugar producers invaded and conquered the world market; the cane sugar producers, who had enjoyed over 90 percent of the world market in the 1840's, had access to less than 50 percent of world demand by the beginning of the twentieth century (Table 7 and Appendix 1).

The sugar interests recognized the threat of European competition, but they were relatively powerless to do anything about it. In 1862, the merchants' Associação Comercial Beneficente de Pernambuco (ACBP) warned that "because of the fall in sugar prices in the major European markets, and the competition with sugar from other producers, who enter those markets with better quality sugars at comparatively lower prices, undoubtedly because of the machinery and tools used in other countries to grow and process this good, we foresee a decline in this very important branch of agriculture, and a decline in provincial income."[4] The planters' Sociedade Auxiliadora da Agricultura de Pernambuco (SAAP) refused to participate in the South American Exposition in Berlin in 1886 because the German beet sugar interests had driven Brazilian sugars off the River Plate markets.[5]

3. "The World's Sugar Production and Consumption," in United States Treasury Department, Bureau of Statistics, Summary of Commerce and Finance of November 1902, pp. 1,267-269.
4. Relatório da Direcção da Associação Commercial Beneficente de Pernambuco apresentado à Assembléa Geral da mesma em 25 de agosto de 1862, p. 5.
5. Sociedade Auxiliadora da Agricultura, "Exposição Sul-Americana de Berlim," Diário de Pernambuco, August 12, 1886. The ACBP and the Associa-

TABLE 7

BRAZIL IN THE WORLD SUGAR MARKET

Years	World Production, Cane and Beet (metric tons)	Beet Sugar/ World Production (percent)	Brazil/ World (percent)	Pernambuco/ World (percent)
1841-45	959,078	5.1	9.3	3.3
1846-50	1,146,281	9.3	10.3	4.2
1851-55	1,433,105	13.7	8.6	4.0
1856-60	1,676,492	21.3	6.3	2.9
1861-65	1,912,388	25.9	6.6	2.4
1866-70	2,414,270	32.0	4.4	2.6
1871-75	3,003,043	40.0	5.7	2.6
1876-80	3,320,512	44.2	5.3	2.8
1881-85	4,333,972	51.2	5.3	2.4
1886-90	5,572,260	56.5	2.8	2.1
1891-95	7,243,020	52.0	2.1	n.a.
1896-1900	8,174,820	61.0	1.4	0.5
1901-05	10,414,020	50.0	0.8	0.1
1906-10	12,831,200	49.3	0.4	0.3

NOTE: Absolute "world" production is unknown. The figures refer to that sugar, mostly exports, for which statistical records exist.

SOURCES: Deerr, *The History of Sugar*, v. II, p. 490-491. "O Açucar na vida econômica do Brasil," pp. 233-235. Our Table 3.

But the most serious attempts to mitigate beet sugar competition resulted not from Brazilian but from international initiatives to eliminate "unfair" competition by producers receiving official subsidies in the form of tax rebates, also called drawbacks or bounties. Conferences including most major western European producers met in 1864, 1868, 1869, 1872, 1873, 1875, 1876, 1887, and 1898, but they reached only minor agreements on methods of determining sugar types. Finally at the Brussels conference of 1901 the European nations agreed to eliminate bounties, limit protective tariffs and preference for colonial manu-

ção Commercial Agrícola, (ACA) did participate. Alguns Agricultores, "A exposição de Berlim," *ibid.*, August 13, 1886. "Provincia de Pernambuco, Exposição Sul-Americana em Berlim," *ibid.*, August 19, 1886.

facturers, and also to discriminate against sugar still enjoying bounties.[6]

Brazilians reacted cautiously to the Brussels convention of 1901. One year before the conference, the First National Agricultural Congress in Rio de Janeiro had recommended imitating the beet sugar nations and returning to exporters a bounty collected at the factory.[7] After the Brussels conference, which Brazil did not attend, the question of adherence confronted sugar interests. Most Pernambuco spokesmen at the Bahia Sugar Conference of 1902 opposed bounties. Those favoring joining the Brussels agreement, including the SAAP, thought such a step would be the only means of preserving Brazil's access to European consumers whose governments supported the agreement. They also insisted that lowering protective tariffs would force the sugar industry to rationalize and improve efficiency. Opponents feared chiefly that if Brazil lowered protective tariff barriers, domestic markets would be invaded by cheaper European beet sugar.[8]

The 1905 Recife Sugar Conference continued debating the question of adherence, and concluded by recommending that the Brazilian government first determine whether Brazil granted direct or indirect bounties, then modify tariffs accordingly to guarantee access to British and other European markets, and

6. Summaries of these sugar conferences can be found in "The World's Sugar Production and Consumption," pp. 1,273-275, and Noel Deerr, *The History of Sugar*, 2 volumes (London, 1949-50) v. II, pp. 505-507. The Brussels agreement is published in the former, pp. 1,361-362.

7. "Congresso de Agricultura—Redação final das conclusões Votadas. Lavoura e Commércio do Assucar e de seus produtos," *Diário de Pernambuco*, October 20, 1901. In Recife, at least one mill owner objected for fear that other countries would impose reciprocal duties. "Sociedade Auxiliadora da Agricultura," *ibid.*, December 10, 1901.

8. Instituto do Açucar e do Álcool, *Congressos Açucareiros no Brasil* (Rio de Janeiro, 1949), p. 114. "Auxiliadora da Agricultura," *Diário de Pernambuco*, April 5, 1902. "Conferencia Assucareira. Estudo da Questão dos Premios," *ibid.*, July 25, 1902. Luiz Correa de Brito, "Secção Agricola. O Brazil e a Conferencia de Bruxellas, I-II," *ibid.*, February 4-5, 1903. Ignacio de Barros Barreto, "A convenção de Bruxellas e o assucar brasileiro," *O Agricultor Pratico* (Recife), June 15, 1903, p. 10. Sociedade Auxiliadora de Agricultura de Pernambuco, "Secção Agricola," *Diário de Pernambuco*, May 6, July 8, 1903. Allan Patterson, "Publicações e Pedido. O Assucar," *ibid.*, November 28, 1904. Serzedelo Corrêa, "A Questão dos Assucares," *ibid.*, December 20, 1904.

make commercial agreements with other countries.[9] Although Brazil adhered briefly to the Brussels agreement, export figures for the period 1906-10 do not denote any substantial benefits.

Beet sugar squeezed Brazil out of European markets. Since the 1840's Brazil had sold most of its sugar to England; the expiration of the Brazilian-British commercial treaty of 1827 had allowed Brazil to send to England sugars competitive with those of the British West Indies, and the passage of the Sugar Act of 1846 had reduced tariffs on crude sugar to favor British manufacturers and consumers.[10] But when beet sugars reached England in the 1870's, Brazil rapidly lost ground (Table 8).

To replace the English markets, Brazilians exported to the only area not yet captured by beet sugar, the United States of America; but even there Brazil's position was tenuous. Sales to the U.S. climbed rapidly in the 1870's, and by the end of the Empire the U.S. had replaced Great Britain as Brazil's principal foreign market (Table 8).

Brazil's new republican government signed a reciprocal trade agreement with the U.S. in 1891 in an effort to improve access to that northern market. But Pernambucans saw little advantage for them in the treaty. The SAAP secretary noted that Brazilian crude sugar already entered the U.S. duty-free, under the terms of the McKinley Tariff of 1890, while the ACBP pessimistically expected that beet sugar would soon replace Brazilian sugar and that the U.S. would sign reciprocal trade treaties with other cane sugar producers, which would end Brazil's advantage.[11]

        9. *Trabalhos da Conferência Assucareira do Recife, (2º do Brasil)* (Recife, 1905). Part 1, pp. 100-119, 127, 133; Part 2, pp. 3, 25-34, 58-68, 126-128, 152.

        10. Alan K. Manchester, *British Preëminence in Brazil: Its Rise and Decline* (Chapel Hill, N.C., 1933), pp. 89, 210, 288. Leslie Bethell, *The Abolition of the Brazilian Slave Trade. Britain, Brazil, and the Slave Trade Question, 1807-1869* (Cambridge, Eng., 1970), pp. 224-228, 273-284. During the five-year period 1841-45, Britain imported an annual average of 863 kilograms of unrefined Brazilian sugar; in 1846 alone, Britain imported 4,972 tons of that sugar. *Parliamentary Papers*, 1852-53, HCC, v. XCIX, *AP*, v. XLIII, pp. 592-593.

        11. Henrique Augusto Milet to Associação Commercial Beneficente de Pernambuco, March 20, 1891; and "O Assucar e seus Derivados," both in *Relatório da Direcção da Associação Commercial Beneficente de Pernambuco apresentado à Assembléa Geral da mesma em 22 de agôsto de 1891*, pp. 79, 20-25.

TABLE 8

PRINCIPAL FOREIGN CUSTOMERS FOR BRAZILIAN SUGAR

| | Great Britain | | U. S. | |
Years[a]	Nominal £	Percent Total Brazilian Sugar Exports	Metric Tons	Percent Total Brazilian Sugar Exports
1850-54			6,808	5.3
1855-59	988,712	40.4	8,365	8.3
1860-64	1,032,270	53.1	5,871	5.2
1865-69	1,253,286	73.0	7,049 [b]	6.5
1870-74	1,795,723	76.3	22,461 [b]	3.3
1875-79	1,830,574	77.8	33,691	20.1
1880-84	1,679,361	63.5	108,282	45.5
1885-89	721,928	47.7	94,977	63.1
1890-94	269,625	12.6	75,162	51.0
1895-99	178,929	13.1	62,859	49.0
1900-04	93,153	11.1		
1905-09	216,023	50.6		

[a] In this table we adopt a slightly different five-year period to conform to available data.

[b] Averaged for three years only.

SOURCES:"O Açucar na vida econômica do Brasil," pp. 234-235. Graham, *Britain and the Onset of Modernization*, p. 75. U.S. Congress, House of Representatives, *Executive Documents*, "Commerce and Navigation of the United States for the Fiscal Year ended June 30, 1850" through ". . . June 30, 1889" (various titles). "The World's Sugar Production and Consumption," pp. 1,384-385.

Subsequent events confirmed these fears. Only months after concluding the treaty with Brazil, the U.S. signed a similar treaty with Spain, to allow Cuban and Puerto Rican raw sugars free entry. Although Brazil later denounced the 1891 treaty, sugar remained on the free entry list of the McKinley Tariff and the Dingley Tariff (1897).[12] As a consequence of its intervention in the Cuban war for independence, the U.S. annexed Puerto Rico and Hawaii and signed a reciprocity treaty with Cuba in

12. E. Bradford Burns, *The Unwritten Alliance. Rio Branco and Brazilian-American Relations* (New York and London, 1966), pp. 60-64. Eduardo Prado, *A Illusão Americana*, 3rd ed. (São Paulo, 1961, first published 1893), pp. 149-151.

1903. These acquisitions created sugar colonies very close to the U.S., which did not sign the Brussels agreement, just as the British West Indies had earlier supplied England. Under a comparable series of "colonial preference" arrangements, the metropolitan market was guaranteed and independent foreign producers like Brazil suffered discrimination.

With the major European countries, England, and the United States either self-sufficient or importing from colonies, Brazil might still have salvaged export markets in Latin America, where transport costs were favorable. But the most populous nations, Mexico and Argentina, as well as Peru, had already developed their own sugar industries by the turn of the century (see Appendix 2). Thus Brazil could find no new markets to replace Britain and the U.S.

Dependency upon a few export markets, whether in England or the United States, entailed other risks besides exclusion. The relatively industrialized nations preferred to import raw or crude sugars which the home refining industry would process for the consumer. As a result, the overwhelming majority of Pernambucan sugars exported were the crude *mascavado* variety. Exporting only mascavado meant earning a price 25 to 33 percent below that paid for refined white *branco* sugar, and paying freight charges for valueless impurities which added bulk and weight. But if they wished to export, producers had no choice over quality; between 1854 and 1874, Britain imposed tariffs on refined sugar imports to protect local refiners. As one English sugar historian has noted: "During the whole three hundred years of the British sugar industry, there has been a clash of the British sugar industry interests between the producer and the refiner, and it is not going too far to say that there has been a tendency to reduce the former to the position of a bond servant to the latter."[13] The U.S. also favored its domestic refineries by retain-

13. Deerr, *The History of Sugar*, v. II, pp. 441-442, 467-468. Compare Deerr's tariff series with that of Geoffrey Fairrie, *Sugar* (Liverpool, 1925), pp. 223-224. The English sugar refining industry's strength did not entail the sugar machinery manufacture's export weakness. While the latter did not ship the most advanced refining machinery to Brazil, they nevertheless did a lively business equipping the engenhos. Graham, *Britain and the Onset of Modernization*, pp. 86-87.

ing duties on refined sugars, in the 1890 tariff and the commercial treaty with Brazil of 1891.[14]

Domestic Sales

The Brazilian domestic market offered the only outlet for producers unable to meet foreign competition. In the Second Empire, Pernambucans sold between 15 and 20 percent of their sugar to home consumers (Table 9). The vast majority of this sugar was a refined branco variety, for three reasons. First, the Brazilian consumer, like the European, demanded white sugar on his table. Second, since imperial Brazil had no large refineries, the planters themselves, or small refiners in Recife, had to transform the mascavado into white sugar. Lastly, the imperial government imposed import taxes on foreign refined and crystallized sugars. While these tariffs were usually designed for fiscal reasons, to raise revenue for the Treasury, they frequently had protectionist effects. Thus during the period 1844-57, imported refined sugars paid 60 percent ad valorem; during the years 1874-81, those sugars paid between 30 and 50 percent. These duties encouraged sugar planters to market their own poorly refined branco to the domestic consumer.[15] Under the republic, on the other hand, the Pernambuco producers apparently fell under the control of large refineries in the center-south, where well over 50 percent of their production was sent, for they were selling almost exclusively mascavado sugar by 1910.

The Brazilian domestic sugar market differed from the international sugar market in one important aspect: given protection from outside competition, the producers could maintain high

14. Those refineries initially enjoyed tariff protection to enable the infant industries to survive. Later, after the formation of the sugar trust, the tariff was used to protect the oligopoly from foreign competition. Alfred S. Eichner, *The Emergence of Oligopoly, Sugar Refining as a Case Study* (Baltimore and London, 1969), pp. 95-97.

15. Nícia Vilela Luz, *A Luta pela Industrialização do Brasil (1808 à 1930)* (São Paulo, 1961) pp. 22n, 36. Under the commercial treaties with the British (1810, 1827) Brazil had been authorized to protect its sugar producers by imposing duties on sugars imported from the British West Indies. Thus, in partial compensation for British exclusion of Brazilian refined sugars, Brazilians protected their own domestic market. Manchester, *British Preëminence*, pp. 89, 208.

TABLE 9

PERNAMBUCO SUGAR SALES TO DOMESTIC MARKET

Years	Average annual quantity (tons)	Average annual value (£ 1880)
1856-60	12,177	310,623
1861-65	10,628	199,813
1866-70	10,484	170,755
1871-75	13,392	198,173
1876-80	17,241	284,226
1881-85	21,581	355,114
1886-89	24,767	497,680
1897-1901	78,016	

SOURCES: *Relatório da Direcção da Associação Commercial Beneficente de Pernambuco apresentado à Assembléa Geral da mesma em 25 de agôsto de 1862. Informações sôbre o Estado de Lavoura* (Rio de Janeiro, 1874), p. 162. *Relatório . . . ACBP . . .8 de agôsto de 1884. Ibid., 9 de agôsto de 1889.* Peres and Peres, *A Industria Assucareira em Pernambuco,* pp. 109-110. José Maria Carneiro da Cunha, "O Assucar," *Diário de Pernambuco,* January 15, 1903. Ónody, *A Inflação Brasileira,* pp. 22-23. Imlah, *Economic Elements in the Pax Britannica,* Table 8.

prices by restricting the supply of their product—that is, demand was relatively price inelastic. In Pernambuco, this idea was first broached for the 1895-96 harvest, when the ACBP and others proposed that the mills manufacture greater amounts of lower priced crude sugar for export, and thereby reduce shipments of white sugar for the home market. They argued that mills producing only export quality crude sugar through October of each year would enjoy freedom from competition by both European beet and Cuban cane sugar, which reached the international market in late October and December, respectively. The 1895-96 scheme failed, however, as did a similar attempt in 1901, because while Pernambuco sent sugar to foreign markets to restrict domestic supply and raise domestic prices, the center-south producers sold in the domestic market and enjoyed the benefit of the northeastern "sacrifice."[16]

16. *Relatório da Direcção da Associação Commercial Beneficente de Pernambuco apresentado à Assembléa Geral da mesma em 9 de dezembro de 1896,* pp 16-17. *Ibid., 9 de agosto de 1897,* pp. 34-38. *Ibid., 13 de agosto de 1901,* p. 11. *Ibid., 12 de agosto de 1902,* pp. 130-143. "Associação Commercial Agricola," *Diário de Pernambuco,* July 8, 1902. Davino Pontual, "Organização agricola e industrial do assucar em Pernambuco," *Boletim da União dos Syndicatos Agri-*

Brazil's adherence to the Brussels agreement of 1902 raised the spectre that if producers drove the price up too high, foreign sugars could enter and take over the domestic market. Consequently a closer co-ordination became necessary. The 1905 Recife Sugar Conference heard a monograph analyzing cartels; various writers began publishing proposals for producers' associations.[17] Encouraged by the sugar factors Mendes Lima e Cia., who financed many of their harvests, the usineiros agreed for the 1906-7 harvest to manufacture a certain percentage of crude *demerara* sugar for export, and a group of warehousers and factors committed themselves not to sell to national markets before November. This cartel, enjoying the support of the SAAP and the União dos Syndicatos Agricolas de Pernambuco (USAP), became known as the Colligação. It was also aided by the Sociedade Nacional da Agricultura (SNA), which issued biweekly statistical bulletins with regional production and consumption data, and assigned export quotas to the various sugar-producting states, in an effort to coordinate their cartels.[18]

The Colligação functioned during the 1906-7 and 1907-8 harvests. While the southern wholesalers boycotted Pernambuco sugar for four months, to drive down prices, the Pernambucans gained the cooperation of producers in Alagoas and Bahia, and even convinced those in Campos, in the state of Rio de Janeiro, to collaborate. Their success led them to plan a national cartel.[19]

colas de Pernambuco (Recife), v. I, no. 3, (March 1907), p. 124. This publication will hereafter be abbreviated *Boletim USAP*.

17. J.G. Pereira de Lima, "Os cartells," *Trabalhos da Conferência Assucareira*, Part 2, pp. 132-146. Luis Correia Britto, "Commercio de Assucar," *Diário de Pernambuco*, August 16, 1905. *Idem.*, "Organização Commercial dos Agricultores," *O Agricultor Prático*. "Auxiliadora da Agricultura," *Diário de Pernambuco*, August 15, December 3, 1905. "O commercio do assucar," *ibid.*, January 20, 1906. J.G. Pereira Lima, "Valorização do Assucar, Memoria apresentada ao Exm. Sr. Dr. Affonso Augusto Moreira Penna," *ibid.*, June 9, 1906.

18. "Valorização do Assucar. A Sociedade Auxiliadora," *Ibid.*, August 25, 1906. "Reunião Agrícola," *A Provincia*, October 5, 1906. "Observações," *Boletim USAP*, v. I, no. 7 (July 1907), p. 551. "Vida Commercial," *Diário de Pernambuco*, August 20, October 29, 1907. "A Colligação e o Decreto do Governado do Estado," *Pernambuco* (Recife), March 1, 1909. "Organização commercial da industria assucareira," *Boletim USAP*, v. I, no. 4, April 1907, pp. 184-185.

19. For the Colligação statutes, see, "Os estatutos da Colligação, O Furo do Diário," *Diário de Pernambuco*, January 28, 1908. Luiz Correia de Britto, "Palestra," *Boletim USAP*, v. II, no. 10, (October 1908), p. 596.

But the cartels depended upon the voluntary cooperation of producers, factors, and warehousers, all with competing interests and in different states, and so they continually threatened to disintegrate. In the 1908-9 harvest, Rio de Janeiro refineries, in the name of consumer protection, resisted the higher prices and bought directly from Campos and Bahian mills. Recife warehousers soon broke ranks and sold sugar south for prices below those awaited by the Colligação. In a vain attempt to avoid the inevitable, José Rufino Bezerra Cavalcante, a federal deputy and sugar broker, persuaded the Pernambuco governor to lower sugar export taxes for producers and raise taxes on sugar sent to domestic markets. But the tax relief was not large enough to allow the producers to circumvent the warehousers, who refused to export.[20]

The Pernambuco Colligação reorganized in March 1909 to allow it to buy enough sugar to keep prices above a minimum level; it would thus act not merely as an advisory group but as an active trader. But the new directors were unable to raise the necessary capital to begin operations, and the planters lost interest. The failure of the Colligação, and Pernambuco's distance from the major consumer markets in the center-south, prevented the planters' from regaining in domestic markets what they had lost abroad. During the years 1900-05, only 23 percent of the sugar received in Rio de Janeiro came from Pernambuco. The states of Rio de Janeiro and Sergipe each sent 30 percent of the capital's sugar.[21]

Combined sales of other cane products, such as rum, molasses, and alcohol, rarely amounted to more than one-tenth the value of sugar exports during the Empire. These products were

20. "Colligação Assucareira," *Diário de Pernambuco*, February 20 and 25, 1909. Luis Correia de Britto, "Colligação Assucareira," *Boletim USAP*, v. III, no. 3 (March 1909), pp. 175-201. This latter article, by the first president of the Colligação, provides a good summary of the cartel's history. "A Colligação . . . ," *Pernambuco*, March 1, 1909.

21. "Publicações a Pedido. Colligação Assucareira de Pernambuco," *Diário de Pernambuco*, March 10, 1909. "Vida Commercial," *ibid.*, September 12, 1909. "Pela Lavoura. A reunião de hontem," *ibid.*, October 19, 1909. "Pela Lavoura. Colligação assucareira," *ibid.*, October 30, 1909. "A valorisação do assucar na Conferencia de Campos," *Boletim USAP*. v. V, no. 7 (July 1912), pp. 485-486. "Valorização do assucar," *Diário de Pernambuco*, June 9, 1906.

not demanded in the international markets because Pernambuco's principal customers, Britain and the United States, satisfied their needs with rum and molasses produced in the U.S., the British West Indies, Cuba, and Puerto Rico.[22]

Some observers hoped that an expanding domestic demand for rum and alcohol would help compensate for weakened sugar markets. In the 1890's, domestic demand for Pernambuco rum swelled impressively as a result of the abolition of slavery, the beginning of mass European immigration to Brazil, and bumper coffee crops in the center-south. In the first years of the twentieth century, Brazilians explored the possible uses of sugar alcohol. Several towns in the interior of Pernambuco installed alcohol lamps; the SNA sponsored an exposition and conference on alcohol in Rio de Janeiro; Recife newspaper editors encouraged the use of alcohol as a substitute for gasoline in internal combustion motors; and a spokesman at the 1905 Recife Sugar Conference urged the use of alcohol to make fruit wines.[23] As a result of these initiatives, alcohol sales at times nearly equaled the value of sugar exports.

Indexes of Stagnation

Sugar remained Pernambuco's primary export, but sugar's contribution to the local economy declined considerably. Annual customs house receipts, a fair indicator of business activity in an area which exported primary products to import finished goods, grew six-fold between the 1840's and the end of the Second Empire. But sugar export receipts only tripled (Table 10). Average annual sugar revenues per capita employed in agriculture

22. *Parliamentary Papers*, 1860-1886, *passim.*, U.S. Congress, House of Representatives, *Executive Documents, passim.*

23. Arthur L.G. Williams, "Report for the Year 1894 on the Trade of the Consular Districts of Pernambuco," *Parliamentary Papers*, 1895, HCC, v. XCVI, *AP*.v. 36, pp. 4-5,23. "Illuminação a Alcool," *Diário de Pernambuco*, September 10, 1903. "Exposição e Congresso do Alcool," *ibid.*, October 20, 1903. "Secção Aríícola, Cia. de Luz e Força Motriz pelo Alcool," *ibid.*, October 22, 1903. "Secção Agrícola, Congresso das Applicações industriaes do alcool," *ibid.*, November 20, 1903. José Rufino Bezerra Cavalcante, "Secção Agrícola, Crise Agrícola, Illuminação a Alcool," *ibid.*, December 5, 1903. "Ainda o Alcool," *ibid.*, December 13, 1903. J.G. Pereira Lima *et al.*, "O alcool e as bebidas artificiaes," *Trabalhos da Conferência Assucareira*, Part 2, pp. 77-97.

TABLE 10

PERNAMBUCO SUGAR AND THE GROSS PROVINCIAL PRODUCT

(in contos)

Years	Average Annual Customs House Receipts	Average Annual Sugar Export Receipts
1836-40		3,347,706
1841-45	1,792,000	4,345,937
1846-50	2,555,648	6,342,517
1851-55	3,807,938	9,322,333
1856-60	5,407,872	10,143,446
1861-65	6,481,012	8,111,525
1866-70	9,591,988	11,108,847
1871-75	10,962,439	11,118,027
1876-80	9,322,528	13,815,198
1881-85	10,519,292	12,617,276
1886-88	10,853,911	14,353,043
1889-93	12,856,981	
1897-98	17,999,644	

SOURCES: *Relatório da Direcção da Associação Commercial Beneficente de Pernambuco apresentado à Assembléa Geral da mesma em 8 de agôsto de 1878. Ibid., 8 de agôsto de 1884. Ibid., 8 de agôsto de 1889.* "Retrospecto Commercial de 1894," *Diário de Pernambuco,* January 15, 1895. Our Table 3.

and livestock-raising declined from £4.3 in 1872 to £1.4 in 1900.[24] This change occurred despite the emigration of some 10,000 slaves, most of whom left sugar plantations, between 1872 and 1881, and despite heavy capital investments in sugar during the 1890's, two developments which should have raised per capita incomes.

Given the export crisis, Pernambucans could draw only slight satisfaction from having improved their position as Brazil's leading sugar exporter. The province's share of average Imperial sugar exports increased from 45 percent in the early 1850's to 53

24. It is unfortunate that the 1890 census for Pernambuco did not identify occupations, which would have permitted evaluating welfare before the period of heavy capital investments in sugar. Per capita shares of sugar income in the province as a whole did rise from £1.34 in 1872 to £2.03 in 1890, indicating an improvement if the proportion employed in sugar production remained constant. But by 1900 that index had fallen to £0.45. See our Table 3, Brazilian censuses of 1872, 1890, and 1900.

percent in the late 1880's. This latter share was twice that of the nearest competitor, Bahia, and nearly six times that of Rio de Janeiro and São Paulo. Moreover, during the years 1903-10, Pernambuco contributed an average of 60 percent of total Brazilian sugar exports; the nearest competitor was Alagoas, with 27 percent.[25]

But in the larger context of world sugar production, Pernambuco exporters had suffered from developments in the later nineteenth century. In 1850 Brazil was the world's third largest cane sugar producer, behind Cuba and the U.S. By the end of the Empire, Java and the Philippines also ranked ahead of Brazil; at the turn of the century, Mauritius and Hawaii had forged ahead; and by 1910, Puerto Rico had surpassed Brazil's best harvests, and Queensland was coming along fast (Appendix 2). This fall in ranking was not offset by the absolute increase in world consumption, for the new demand did not buy Brazilian sugar. In the 1840's Brazil had supplied better than 9 percent of the world market, and Pernambuco one-third of that share. By the turn of the century, Brazil's share had declined to under 2 percent and Pernambuco to only one-fourth of that percentage (Table 7).

The Brazilian planters, and particularly the Pernambucans, could attribute their difficulties to relatively few causes. The export crisis was a clear product of the beet sugar boom, which drove down prices and preempted traditional markets. Brazil was unable to replace its former European markets with Western Hemisphere outlets because the consumers found preferable alternative supplies, either in colonies or within their own boundaries. Only the protected domestic market remained, and here, because of Pernambuco's distance from the major population centers, the northeastern producers were unable to maintain the ascendancy enjoyed in foreign markets. Without markets, the industry stagnated.

25. Júlio Brandão Sobrinho, *A Lavoura da Canna e a Indústria Assucareira dos Estados Paulista e Fluminense. Campos e Macahé em confronto com S. Paulo. Relatório apresentado ao Illm. e Exm. Sr. Dr. Antônio de Pádua Salles* (São Paulo, 1912), p. 100. Gileno Dé Carli, "Alagôas, Sinópse Histórica do Açucar," *Annuário Açucareiro para 1935*, p. 38. Diógenes Caldas, "Parahiba, Sinópse Histórica do Açucar," *Annuário Açucareiro para 1935*, p. 108; Eudes de Souza Leão Pinto, *Cana-de-açucar* (Rio de Janeiro, 1965), p. 77. *Economia e Agricultura*, ano I, no. 5 (February 15, 1933), p.37.

3

TECHNOLOGICAL PROGRESS

 $\mathscr{T}$ he technology in both the agricultural and the industrial phases of Pernambuco's mid-nineteenth century sugar economy did not use the most modern methods. Some have blamed this backwardness for the sugar industry's stagnation.[1] But a minority of planters converted to more modern technology after 1870. If the industry as a whole did not make new investments in more productive technology, various factors, including the industry's stagnation itself, could be blamed. A review of cane agriculture, sugar manufacture, and transportation exposes certain areas of backwardness; but it also reveals successive innovations by a prosperous sector of the industry, and thereby rebuts the charge of unrelieved obsolescence.

Agricultural Operations

Sugar cane agriculture most clearly reflected Pernambuco's backwardness. The first cane brought to Pernambuco in the sixteenth century, known as Creole cane, predominated for over 250 years until the beginning of the nineteenth century, when King João VI of Portugal captured French Guiana in 1808. Two years later the Portuguese governor of the captured colony sent samples of the Otaiti or Bourbon cane, native of French Tahiti and grown in French Guiana, to the Olinda Botanical Garden in Pernambuco, where they were reproduced and distributed to *senhores de*

1. See, for example, Luis Amaral, *História Geral da Agricultura Brasileira no Tríplice Aspecto Político Social-Econômico*, 2nd ed., 2 vols. (São Paulo, 1958, first published 1940-41), v. I, p. 358

engenho (planters, mill owners). This new variety became known as Cayenne cane after the capital of French Guiana; by virtue of its larger size, greater branching, richer sugar content, and drought resistance, it quickly replaced the Creole cane and remained the dominant variety along with a related species, Imperial cane, for most of the nineteenth century. Other species from Cayenne, Java, and Mauritius were introduced after 1854.[2]

The principal innovations in cane species occurred after 1879. In that year, the cane disease called *moléstia* or cane rot attacked the previously healthy Cayenne cane. To combat moléstia, which left a yellow soluble mucilage in the sugar boilers, whose removal interrupted the manufacturing operation, new cane species were introduced from Java and Mauritius.[3] In 1892 Manuel Cavalcante de Albuquerque, owner of the Engenho Cachoeirinha in Escada, successfully sexually reproduced cane, thereby eliminating the reliance on imported cane sections and making selective hybridization possible.

2. Henry Koster, *Travels in Brazil*, 2 vols. (Philadelphia, 1817), v. II. p. 113. F.A. Pereira da Costa, "Origens Históricas da Indústria Assucareira em Pernambuco," in *Trabalhos da Conferência Assucareira*, pp. xxiv-xxv, xlvi-xlviii. Alinio de Salles, "O Verdadeiro Responsável pela Introdução da 'Cana Caiana' em Nosso País," *Brasil Açucareiro* (Rio de Janeiro) v. LXXII, no. 4 (October 1968), pp. 46-48. For an excellent description of the successive introduction of new cane varieties, and other technical improvements in nineteenth-century sugar manufacture in Brazil, see Alice P. Canabrava, "A Grande Lavoura," in Sérgio Buarque de Holanda and Pedro Moacyr Campos (editors), *História Geral da Civilizaçao Brasileira*, 6 vols. (São Paulo, 1963-71) tomo II, *O Brasil Monárquico*, v. 4, *Declínio e Queda do Império* pp. 102-110.

3. The disease had first appeared in Cabo under the name "cane gummosis." Gileno Dé Carli, "Geografia Econômica e Social da Canna de Açúcar no Brasil," *Brasil Açucareiro*, v. X, no. 2 (October 1937), p. 140. *Relatório com que o Exm. Sr. Dr. Adolpho de Barros Cavalcanti de Lacerda Passou ao Exm. Sr. Dr. Adelino Antônio de Luna Freire primeiro vice-presidente a administração desta província em 18 de setembro de 1879*, p. 24. *Relatório com que o Exm. Sr. Dr. Adelino Antônio de Luna Freire passou ao Exm. Sr. Dr. Lourenzo Cavalcante de Albuquerque a administração da província em 29 de dezembro de 1879*, p. 29. "Negocios da Lavoura," *Diário de Pernambuco*, March 19, 1880. SAAP, Livro de Atas no. 1, March 25, 1882. Report by Paulo d'Amorim Salgado in Henri Raffard, *O Centro da Indústria e Commércio de Assucar no Rio de Janeiro* (Rio de Janeiro, 1892), p. 109. "A molestia da canna," *O Brazil Agricola, Industrial, Commercial, Scientífico, Litterario, e Noticioso* (Recife), 2nd series, anno I, no. 2 (September 12, 1879), pp. 4-5. SAAP, Livro de Atas no. 2, (Conselho Administrativo), December 13, 1881.

Cavalcante de Albuquerque himself created the Manteiga variety.[4]

Cane growing was highly labor-intensive.[5] In July, August, and early September for the hilly lands, and between September and November for the rich *várzeas* or riverside lands, the planters sent slave gangs with hoes out to the fields to plant short pieces of cane in furrows. Within two weeks, the cane joints sprouted roots, leaves, and other buds which also took root and sprouted. Slaves with hoes and brush hooks loosened the soil after five weeks, and cleared away weeds and the lower leaves from the cane stalks two or three times. The plow, harrow, and cultivator were not generally used, although sugar planters in Louisiana and Cuba employed such equipment in the 1840's. In Pernambuco planters used crude ox-drawn plows on the várzeas; by the late 1880's, they were importing North American, German, and French plows, but they still relied on the hoe for furrowing the hillsides, allegedly because tree stumps and roots made plowing

4. José M. C. da Cunha, "Canna de Assucar e suas diversas sementes selecionadas, seus resultados em Campo de Experiência, commercio de assucar e alcool," *Trabalhos da Conferência Assucareira*, Part 2, pp. 8-11. Henrique Augusto Milet, "Cannas oriundas das sementes de flexa da canna," *Diário de Pernambuco*, October 20, 1892. *Idem.*, "Agricultura, SAAP em 1 de Maio de 1893," May 3, 1893. "Cultura de Cannas no Municipio da Escada," *Boletim USAP*, v. II, no. 2 (February 1908), pp. 657-700.

5. For descriptions of sugar cane agriculture and sugar manufacture in early nineteenth-century Pernambuco, see Koster, *Travels in Brazil*, v. II, pp. 105-126, and James Henderson, *A History of the Brazil, Comprising its Geography, Commerce, Colonization, Aboriginal Inhabitants, etc., etc., etc.* (London, 1821) pp. 391-392. For accounts in the latter part of the century, see F.L.C. Burlamaque, *Monographia da Canna d'Assucar* (Rio de Janeiro, 1862), pp. 73-227; speech by Sr. Nabor, *Annaes da Assembléa Provincial de Pernambuco, sessão de 1864* (Recife, 1870) Tomo I, April 9, 1864, pp. 444-447; Herbert H. Smith, *Brazil: The Amazons and the Coast* (New York, 1879), pp. 445-447; and Raffard, *O Centro da Industria*, pp. 111-117. Unless otherwise noted, our discussion synthesizes these sources. Other descriptions can be found in Manuel Diégues Júnior, *População e Açúcar no Nordeste do Brasil* (São Paulo, 1954), pp. 131-149; and Philarete Carneiro Nobre de Lacerda, "Fabricação de Açúcar Bangue nos Aureos Tempos de 1890 à 1910," *Brasil Açucareiro*, ano XXXVII, v. LXXIV, no. 2 (August 1969), pp. 123-129. These latter two sources do not refer explicitly to techniques used in Pernambuco.

impractical. Recommendations during the moléstia to plow deeper to turn up more resistant soils went unheeded.[6]

Cane matured in twelve to fifteen months. The harvest began when the rainy season ended, usually between late August and October. Slaves with machetes and brush hooks cut and stacked the cane in bundles of six to twelve canes on horses, mules, and oxcarts for conveyance to the mill.[7] The most significant mechanization of the agricultural phase of sugar production occurred in cane transport. The 1870's witnessed the first private railroads installed between canefields and engenhos to carry larger quantities of cane more rapidly to the mill. While these early railroads were horse-drawn affairs on portable track, later installations were permanent. Nearly all the usinas installed small steam-powered railroads extending as far as thirty kilometers on gauges varying between 0.60 and 1 meter. Some of those railroads built with state government subsidies, such as those at the Usinas Cachoeira Lisa and Bom Fim, not only trafficked between the fields and the mill but also connected with the Recife and San Francisco Railroad, and were obliged to carry public freight and passengers.[8]

Negligible costs of forests and land led the planters to abuse

6. J. Carlyle Sitterson, *Sugar Country: The Cane Industry in the South, 1753-1950* (Lexington, Ky., 1953), pp. 115, 128. Manuel Moreno Fraginals, *El Ingenio: El Complejo Económico Social Cubano del Azúcar, Tomo I (1760-1860)* (La Habana, 1964), pp. 91-92. *Relatório apresentado à Assembléa Geral na terceira sessão da décima sétima legislatura pelo Ministro e Secretário do Estado dos Negócios da Agricultura, Commercio e Obras Públicas Pedro Luiz Pereira de Souza* (Rio de Janeiro, 1881), pp. 3-6.

7. Theodoro Cabral, "Vocabulário Açucareiro de Antonil," *Brasil Açucareiro*, anno III, no. 4 (June 1935), reprinted in Simonsen, *História Econômica do Brasil*, pp. 121-123, found that the bundle *(feixe)* contained twelve canes in the seventeenth century. Henrique Augusto Milet, *A Lavoura da Canna de Assucar* (Pernambuco, 1881), p. 108, quotes one planter who used feixes of six canes. In 1967 I learned that the feixe had risen to twenty canes; since cane cutters earned by the feixe, the increase represented a speedup.

8. "'Industria agricola assucareira em Pernambuco, *Diário de Pernambuco*, July 21-22, 1881. Um visitante, "Agricultura," *ibid.*, March 31, 1881. *Mensagem apresentada ao Congresso Legislativo do Estado em 6 de Março de 1899 pelo Governador Dr. Joaquim Corrêa de Araujo*, p. 69.

wood and soil resources. Originally, the planters grew cane in the várzeas' rich black *massapê* soil, famed for fertility, easy cultivation, and good water retention. By the nineteenth century, however, these várzeas were mostly in use, and the planters were deforesting hillsides. The hillside soils, not as soft or porous as the várzea soils, were harder to work and the slaves often had to leave tree stumps in the ground. Without terracing, contour plowing, or reforestation, this deforestation accelerated soil destruction and impeded natural irrigation. Modernization of the sugar industry only compounded the problem, because the large usinas required even greater cleared areas for cane and much more firewood to heat the boilers.[9]

The fertile várzeas allowed cane to spring from the stubble ratoons in the second, third, and even fourth years after planting, although sugar quality declined proportionately. After the fourth year, the most common practice was to let the land lie fallow, rather than rotating food crops. Although they left cane leaves and cut weeds in the fields, planters normally did not fertilize. Even manuring by grazing animals was not practiced. Only during the moléstia plague did the SAAP import guano, sodium nitrate, and superphosphate to strengthen the poor soils blamed for the disease. But once the threat passed, the cure was forgotten as well.[10]

9. Nemo, "A Associação Beneficente de Pernambuco e as causas da decadencia agricola," *Diário de Pernambuco*, January 28, 1875. *Relatório* by Louis Leger Vauthier dated 1844, cited in F.A. Pereira da Costa, "Codigo Florestal," *Trabalhos da Conferência Assucareira*, Part 2, pp. 76-77. Pereira da Costa also mentioned another French public works director in 1863, and an English engineer in 1884. Both experts agreed substantially with Vauthier. "Interesses Provinciaes," *O Progresso* (Recife), Tomo II (1847), pp. 706-708. José Antonio de Almeida Pernambuco, "Governo do Estado de Pernambuco," *Diário de Pernambuco*, September 28, 1895. SAAP, *Acta da Sessão Solemne da Assembléa Geral de 28 de setembro de 1882 e Relatório Annual do Gerente Ignacio de Barros Barreto* (Recife, 1882), p. 114. "O Combustivel na industria assucareira," *O Industrial* (Recife), Anno I, no. 10, (October 15, 1883), p. 111.

10. Report by Doyle, *Parliamentary Papers*, 1872, HCC, v. LVIII, AP, XXIII, p.634. SAAP, *Acta de 28 de Setembro de 1882 e Relatório*, pp. 26, 31, 73-75, 83-85, 95-99. *O Brazil Agricola*, Anno III, no. 10 (May 31, 1882), p. 143. "A canna de assucar em Pernambuco," *Diário de Pernambuco*, March 21, 1882. "The CSFB," *ibid.*, June 18, 1885.

Industrial Operations

The sugar manufacturing, a more capital-intensive operation, traditionally began when large hardwood rollers covered
with iron hoops or plates crushed the cane to extract its juice,
which constituted 90 percent of cane weight. By the mid-nineteenth century, Pernambucan manufacturers had generally converted from vertical rollers to horizontal rollers which allowed
greater pressure, faster grinding, and better distribution of the
cane across the width of the rollers. By the end of the century,
cast iron rollers had replaced wooden ones.

Pernambucans did not adopt diffusion, the principal alternative method of extracting juice from cane. Invented by the beet
sugar manufacturers, diffusion eliminated the grinding rollers
and substituted knives which split the cane lengthwise. The
cane was then repeatedly immersed in water to wash out the
syrup. Diffusion allegedly extracted a greater percentage of
purer syrup from the cane, but it also required a larger amount of
fuel to evaporate the water, and frequent honing of the slicers. In
Pernambuco, only two mills, the Usinas Ipojuca and Cucaú,
installed diffusion machinery. The former's experience was unsa-
tisfactory, and the owners soon reconverted to grinding mills. The
latter used diffusion successfully, and partially solved the fuel
problem by installing a special furnace to dry bagasse for burning
and by experimenting with petroleum. When the owners
expanded capacity, however, they purchased a grinding mill to
work alongside the diffusion apparatus.[11]

Power sources in the mills underwent considerable changes
in the nineteenth century. In the early years, teams of horses,
oxen, or mules were harnessed to long levers attached to the roll-

11. F.M. Draenert, "A diffusão no engenho central de Barcellos," *ibid.*,
October 15, November 27, December 23, 1886. Henrique Augusto Milet and
others criticized Draenert and diffusion severely, *ibid.*, September 17, October
29, December 2, 1886; October 19, 1887; December 28, 1888; and September
7, 1893. Emilio Dolé "Usina Cucaú," *ibid.*, February 18, 1895. Luiz de Castil-
hos, "Industria assucareira," *ibid.*, November 7, 1893. "Usina Cucaú," *A Provincia*, February 12-13, 1895. "A diffusão," *Diário de Pernambuco*, August 25,
1895. Peres and Peres, *A Industria Assucareira em Pernambuco*, pp. 136-140.

er axles, in a fashion little changed since the sixteenth century. As late as 1854, a survey in Pernambuco reported that 426 out of 532 mills, or 80 percent, still relied on animal power, and an 1857 count showed 66 percent of 1,106 mills used animals for power. Mills near rivers used water wheels to turn rollers: in the 1854 and 1857 surveys, water power turned 19 and 31 percent of the mills counted. Steam engines, on the other hand, operated in only 1 and 2 percent of the mills, respectively, although in Louisiana steam-driven mills outnumbered animal-powered mills by 1840, and in Cuba 70 percent of the 1,358 mills counted in 1860 used steam.[12]

After 1870, mill owners introduced steam engines more rapidly. By 1871, 6 percent of the 440 mills shipping sugar via the Recife and San Francisco Railway used steam power; by 1881 the comparable figure was 21.5 percent of 609 mills, and by 1914, 34 percent of 2,288 engenhos in the state used steam, as well as all 62 modern usinas.[13]

After extraction the juice was boiled to evaporate the water, which was 70 to 80 percent of juice content, in one or two series of four to seven boiling kettles, each with its own furnace, Copper kettles, lighter and better heat-conductors, were preferred to the cheaper but rust-prone iron pots, and during the nineteenth century a more economic central furnace was placed beneath the first kettle with flues carrying the heat to the rest of the kettles. This system was known as the Jamaican train. Wood fueled the furnace which heated the kettles. Mill owners generally did not

12. Simonsen, *História Econômica do Brasil*, p. 99. Fernando de Azevedo, *Canaviais e Engenhos na Vida Política do Brasil* (Rio de Janeiro, 1948), pp. 210-211. *Relatório que à Assembléa Legislativa Provincial de Pernambuco apresentou no dia da abertura da sessão ordinaria de 1854 o Exm. Conselheiro José Bento da Cunha e Figueiredo, Presidente da mesma Provincia*, Table. *Relatório que à Asssembléa Legislativa Provincial de Pernambuco apresentou no dia da abertura da sessão ordinaria de 1857 o Exm. Sr. Conselheiro Sérgio Teixeira de Macedo*, p. 75. Sitterson, *Sugar Country*, p. 138. Ramón de la Sagra, *Cuba, 1860: Selección de artículos sobre agricultura cubana* (La Habana, 1963), p. 138.

13. Recife and San Francisco Railway Company (Limited), *Report of the Proceedings at the Fifty-Fourth Half-Yearly Ordinary General Meeting of the Shareholders . . . June 30, 1882*, p. 2. Peres and Peres, *A Industria Assucareira em Pernambuco*, pp. 32-33.

burn bagasse, although the practice was known since 1812. One authority specifically warned that "to burn bagasse is to steal from oneself, since if this bagasse were buried in the canefields, it would fertilize them."[14]

To the boiling kettles, the sugar master added wood ashes to remove soluble impurities and whiten the sugar. Other clarifying agents were burnt animal bones and blood, lime, and sulphuric acid. The soluble impurities floated to the top of the kettles as a foam, which was skimmed off and used to make rum, to feed livestock, returned to the beginning of the kettle series for reprocessing, or discarded. Non-soluble impurities in the cane juice were removed by passing the cane juice through linen or wool cloth filters and metal strainers as it was moved from one kettle to the next.

In the later Empire, a few mill owners introduced the vacuum pan, invented in 1813, for evaporating. This closed pot lowered the boiling temperature and thereby economized fuel and reduced the risk of carmelization. The multiple—or triple—effect vacuum pan economized even further, by connecting the boilers so that one boiler's vapor would heat a second boiler, whose vapor in turn would be condensed to make new steam. Vacuum pans appeared in Cuba, Peru, Mexico, and Louisiana in the 1840's, and were fairly common in Cuba by 1860, but Pernambucans did not begin to install them until the 1870's.[15] Steam heat represented another advance in the boiling phase. Instead of an open flame on the kettle, steam-heated pipes, spheres, or serpentines immersed in the juice, or a double-bottomed pan, heated and evaporated more efficiently.

When the juice reached a syrupy consistency, it was poured into pots for cooling and crystallizing. During the first 24 hours,

14. Burlamaque, *Monographia da Canna d'Assucar*, p. 70. José Honório Rodrigues, "A Literatura Brasileira sôbre Açucar no Século XIX," *Brasil Açucareiro*, v. XIX, no. 5 (May, 1942), pp. 468-469. José Wanderley de Araujo Pinho, *História de um Engenho do Recôncavo Matoim-Novo Caboto Freguezia, 1552-1944* (Rio de Janeiro, 1946), pp. 157-158.

15. Noel Deerr, *Cane Sugar: A Textbook on the Agriculture of the Sugar Cane, the Manufacture of Cane Sugar, and the Analysis of Sugar House Products*, 2nd ed. (London, 1921), pp. 608-10. *Idem.*, *A History of Sugar*, v. II, p. 561. Sitterson, *Sugar Country*, p. 147. Moreno Fraginals, *El Ingenio*, p. 119.

many sugar crystals rose to the top of the cooling syrup and could be skimmed off. The rest of the syrup was poured into conical clay or wood forms to cool for four or five days. Then the purging process occurred: water and clay were sprinkled on the top of the sugar loaf, and as these materials percolated downward over six to eight days they extracted the water remaining between the sugar crystals.

The purged loaves dried in the sun for 18 to 22 days, or on racks in large heated barns, until no more molasses drained from a hole in the form's bottom. The dry loaf contained white sugar at the top, yellowish sugar in the middle and brown sugar on the bottom. Centrifuge crystallizers invented in 1837 accelerated this time-consuming process greatly. The cooled syrup was poured into wheels with strainers along the rims. Rapid rotation spun the molasses out the holes and left the sugar crystal on the inside surface of the strainers, which were then washed with water and steam to whiten the sugar. This process, excluding the drying, could produce 60 kilograms of sugar every twenty minutes. A locally manufactured wooden centrifuge was reported in Pernambuco in the early 1850's, and the first imported centrifuge was installed in 1872.[16]

The crude manufacturing process of the earlier nineteenth century allowed the average engenho to produce at best between one and 1.5 tons of mostly mascavado sugar per day—that is, between fifteen and twenty loaves of 60 to 70 kilograms each, or up to 150 tons per harvest. In contrast, the fully equipped usinas at the turn of the twentieth century produced upwards of ten tons per day, or nearly one thousand tons per harvest, including mostly branco sugar.[17]

16. Deerr, *Cane Sugar*, pp. 608-610. William Hadfield, *Brazil, the River Plate and the Falkland Islands* (London, 1854), p. 104. "Relatório da Comissão Directora da Exposição Provincial de Pernambuco em 1872," *Diário de Pernambuco*, February 21, 1873.

17. *Relatório . . . 1854 . . . José Bento da Cunha e Figueiredo*, Table. *Relatório . . . 1857 . . . Sérgio Teixeira de Macedo*, pp. 75-77. *Mensagem apresentada ao Congresso Legislativo do Estado em 6 de março 1900 pelo Exm. Sr. Desembargador Sigismundo Antonio Gonçalves, vice-presidente do Senado no exercício do cargo de Governador do Estado*, pp. 43-45.

In addition to making crude sugar, the mill owner also produced a low grade rum from foams skimmed off the boiling kettles
and the last batches of syrup. He aged the mixture briefly in ear-
then pots, with boiler kettle waste as a fermenting agent, and
then distilled it once. The pipes and barrels of rum reaching
Recife in the earlier nineteenth century contained 80 percent
water. Mills specializing in rum used better-quality stills and
distilled several times; at century's end some usinas devoted
their distilleries exclusively to the manufacture of alcohol.[18]

In Recife, a small number of refineries melted raw sugar
and clarified again. The refiners pressed the mascavado in sacks
for five or six hours, rolled the extract, and then pressed again for
twice as much time. By these chemical and physical processes
they changed mascavado sugar into higher-priced branco sugar,
which found its outlet in the domestic market. Even when the
industry dedicated itself to this domestic market, after the mid-
1890's, the local refineries remained small, for the usineiros
either managed to produce a branco sugar acceptable to the consumers or they sold to large refineries near major markets.[19]

The Pace of Progress

The relatively slow adoption of technological developments
by the Brazilian sugar industry has been explained in several
ways. Some writers have rightly pointed to the abundance of vir-
gin land which encouraged a land-extensive technology in agriculture and militated against soil conservation and improvement.[20] The same historians have also ascribed technological
backwardness to the institution of slavery, for various reasons.
First, slavery provided a relatively abundant and cheap labor
supply, and thereby encouraged a labor-extensive technology in

18. Ferreira Soares, *Notas Estatisticas*, pp. 102-103.
19. Antonio Carlos de Arruda Beltrão, *A Lavoura da Canna e A Industria
Assucareira Nacional* (Rio de Janeiro, 1918), p. 20. Arruda Beltrão built the
short-lived Beltrão refinery in Recife.
20. Prado Júnior, *História Económica do Brasil*, pp. 91-92, Buescu and
Tapajós, *História do Desenvolvimento Económico do Brasil*, pp. 152-153. See
also Koster, *Travels in Brazil*, v. II, pp. 117-118.

both field and mill. Secondly, the senhores de engenho were reluctant to teach more advanced technologies, in part because they themselves were ignorant, but also in part because they feared placing valuable machinery in indifferent or hostile hands. Free labor, on the other hand, by fault of limited educational opportunities, was little better equipped to operate and maintain modern machinery.[21]

The combination of cheap land and cheap uneducated labor produced a conservative attitude of "routinism" toward technological innovation. The English traveler and planter Henry Koster noted that the planters worked "year after year without any wish to improve, and without, indeed, the knowledge that any improvement could be made." The traveling French merchant L. F. Tollenare heard "sarcastic remarks" about planters "who think about the theoretical part of agriculture and make some attempts at improving the cultivation and manufacturing processes." In the 1860's a provincial deputy lamented "this aversion to innovations, this strong attachment to ancient practices, this inveterate routine," and a U.S. Civil War refugee planter from Louisiana described the Pernambuco sugar manufacture as "barbarous." A French visitor reported as late as 1912 that "innovations are introduced with great difficulty and delay." Sometimes, moreover, unscrupulous or incompetent manu-

21. Paul Singer, *Desenvolvimento Econômico e Evolução Urbana* (São Paulo, 1968), pp. 289-290. Prado Júnior, *História Econômica do Brasil*, p. 90. Buescu and Tapajós, *História do Desenvolvimento Econômico do Brasil*, p. 153. Opponents of diffusion argued that the available labor force could not handle the machinery. Henrique Augusto Milet, "A diffusão do assucar da canna," *Diário de Pernambuco*, October 20, 1886. Raul Prebisch has suggested that large landowners may also have been averse to technology which disemployed large numbers of free workers and thereby increased the social threat. Interview, New Brunswick, New Jersey, May 20, 1971. In a slave society, however, this threat could be minimized by the relatively costless reemployment of slaves in other activities, or by selling slaves to more labor-intensive producers. Gilberto Freyre's observation that slaves developed skills as iron miners, metal workers, cooks, "musicians, circus acrobats, bloodletters, dentists, barbers, and even teachers," belies any belief that slaves could not learn higher skills. *The Masters and the Slaves (Casa Grande e Senzala). A Study in the Development of Brazilian Civilization*, translated by Samuel Putnam, 2nd ed. rev. (New York, 1956, first published 1933), pp. 310, 411.

facturers' representatives and defective machinery discouraged individuals initially interested in modernization, and thereby confirmed the others in their routinism.[22]

Cheap land, cheap labor, and "routinism" help explain technological backwardness. Cost was also a prime concern in the adoption of modern technology: most mill owners simply could not afford to buy the equipment. The era of mill modernization began in the early 1870's, when steam engines, vacuum pans, and centrifuges were introduced. In these years, the cost of slaves and free laborers reached the highest levels of the century, and undoubtedly one can attribute the timing of modernization to these high labor costs (see Chapters Seven and Eight). But labor costs alone did not cause modernization, for the heaviest capital investments occurred in the 1880's and 1890's, when labor costs were falling.

Capital costs and market insecurity also affected the rate of innovation. The slow accumulation of retained earnings in the traditional engenho, in comparison with the cost of acquiring modern machinery, prevented the average senhor de engenho from adopting modern technology. Later nineteenth-century data indicate that the profit rates in traditional engenhos did not exceed 9 percent (Table 11).[23] These profits hardly allowed the mill owner

22. Azevedo, Canaviais e Engenhos, p. 216; Prado Júnior, História Econômica do Brasil, p. 90. Koster, Travels in Brazil, v. II, p. 105. L. F. Tollenare, Notas Dominicais tomadas durante uma residência em Portugal e no Brazil nos annos de 1816, 1817, e 1818, Parte relativa a Pernambuco, translated by Alfredo de Carvalho (Recife, 1905), p. 85. Speech by Sr. Nabor in Annaes da Assembléa Provincial, April 9, 1864. Thomas Adamson, Jr., to William Seward, Pernambuco, August 29, 1866, in U.S. National Archives, Despatches from United States Consuls in Pernambuco, 1817-1906, Microcopy T344, v. 8, his emphasis. Paul Walle, Au Brésil du Rio São Francisco à l'Amazone (Paris, 1912), p. 159. Palladius, "Os novos apparelhos de fabricar assucar," Jornal do Recife, August 19, 1881. CP, "Os novos apparelhos de fabricar assucar," Diário de Pernambuco, September 19, 1881. "Parte Industrial e Agricola, I," ibid., August 22, 1893.

23. We cannot take seriously data published by Francisco do Rêgo Barros de Lacerda for his Engenho São Francisco da Várzea: before modernization he reported an incredible profit of 33 contos or 64 percent. He also gave several conflicting estimates for production in 1880-81. "Industria Agricola em Pernambuco," Diário de Pernambuco, July 22, 1881: Francisco do Rêgo Barros de Lac-

to amortize borrowed investment capital at minimum rates of 10 or 12 percent per year. New entry into the industry with modern technology was even more difficult: to mount a complete usina capable of producing 750 tons of sugar per harvest, 135 contos were needed. Diffusion machinery cost considerably less, only 55 contos to mount a mill producing 1,500 tons of sugar per harvest, but the process found few admirers. [24]

The contrast between low profits and high costs for modern equipment does not by itself explain why most Pernambucan senhores de engenho did not modernize in the later nineteenth century. Had they invested in modern machinery, they theoretically could have earned profits so much larger that they could have paid off the investment in ten years while at the same time enjoying additional profits (Table 12).

Despite these possible rates of return, most planters still did not invest in modern equipment. Two economic factors explain this apparently irrational behavior. First, the general insecurity of the world sugar market, over whose prices Brazilian producers had no control, increased the risk of new investments. A few bad harvests, further falls in sugar prices, or new losses of foreign customers, could ruin the enterprising senhor de engenho. Secondly, many planters did not make modernizing investments for lack of adequate credit. Borrowing fifty or one hundred contos or more to be repaid over ten years was quite different from borrowing ten or twenty contos to be repaid at harvest's end, and the prohibitive interest rates charged on long-term capital loans reflected this difference. Indeed even the 10 or 12 percent annual interest paid by José Bezerra and Presciano was low for the period (see Chapter Four). [25]

erda, "Ao Sr. Palladius," *ibid.*, September 14, 1881; "Safras do engenho São Francisco da Várzea,"*ibid.*, August 24, 1886. We also omit calculations by Milet for his own mill, because the cost allocation was unclear. Milet, *A Lavoura da Canna de Assucar*, pp. 7-10. *Idem.*, *Os Quebra Kilos, e a Crise da Lavoura*(Recife, 1876), pp. 4-8.

24. Emilio Dolé, "Engenhos Centraes," *Diário de Pernambuco*, September 27, 1890. "Sôbre Usinas de Assucar," *ibid.*, October 2, 1890. In changing daily into yearly capacities, we have assumed a 100-day harvest. "Engenhos Centraes," *ibid.*, October 19, 1890.

25. For complaints by planters whose investments in modern machinery

TABLE 11
ENGENHO BUDGETS

Costs, Receipts, Profits	José Bezerra de Barros Cavalcanti Revised by H.A. Milet (1876)	Presciano de Barros Accioli Lins (1876)
Costs		
Cane Costs		
Preparing soil, planting, cultivating	6:000$000	6:000$000
Cutting, Transporting	2:700$000	4:000$000
Mill Costs		
Sugar	2:370$000	4:600$000
Rum	226$800	
General Costs		
Rent	2:000$000	3:000$000
Maintenance	2:800$000	2:500$000
Salaries	1:000$000	3:250$000
Interest [a]		
10% on 5:000$000	500$000	
12% on 23:350$000		2:760$000
Total Operating Costs	16:796$000	26:110$000
Receipts [b]		
Sugar		
194 tons	16:852$400	
169 tons		23:920$000
Rum		
6,284 liters	1:408$800	
24,376 liters		3:520$000
Total Receipts	18:261$200	27:440$600
Net Profit [c]	1:465$200	1:680$600
	(8.7%)	(6.4%)

[a] Although identified only as interest, these figures presumably include amortization as well. If not, the capital cost would be higher and the profit rates lower.

[b] With freight and marketing expenses deducted.

[c] Profit rates equal net profits as a percentage of total costs. Capital investment figures, the standard denominator, were unavailable. Undoubtedly the initial investment exceeded the operating costs, however, and the real profit rate would be even lower.   *(Continued)*

SOURCES: Originally published in the *Jornal do Recife*, May 10 and July 28, 1876, reprinted and criticized in Milet, *A Lavoura da Canna de Assucar*, pp. 17-26, 104-112, and reprinted without comment in Peres and Peres, *A Industria Assucareira em Pernambuco*, pp. 230-238.

TABLE 12

RETURNS ON INVESTMENTS IN MODERN TECHNOLOGY

	Engenho Mameluco	Usina Equipped with Mariolle Pinguet Machinery [a]
Increased Production [b]		
Sugar, Branco	57:960$000	102:672$000
Sugar, Mascavado	15:386$000	27:255$200
Total	73:346$000	129:927$200
Increased Costs		
Capital [c]	10:000$000 [d]	30:000$000
Operating [e]	45:371$578	80:363$822
Total	55:371$578	110:363$822
Increased Profits	17:974$422	19:563$378
Increased Profits/ Increased Costs	32.5 %	17.7%

[a] To avoid exaggeration, we assumed Mariolle Pinguet's smallest usina (750 tons of sugar per year) at its highest price (150 contos).

[b] Assuming data for the 1880-81 harvest: branco price, $230 per kilogram; mascavado price, $157 per kilogram. Mameluco produced 72 percent branco and 28 percent mascavado in a total of 250 tons made in 1877-78. We compare that to the 600 tons produced in 1880-81. We assume that the usina before modernization made the customary 130 tons, and in the same branco-mascavado proportion.

[c] All old machinery completely amortized; new machinery 10 percent amortization, 10 percent annual interest.

[d] Assuming Antônio Marques spent 50 contos on Mameluco's machinery, as did Rêgo Barros and the Barão de Muribeca for their mills.

[e] Operating costs for both mills estimated at 8$218 per loaf with one-half the cane supplied by sharecroppers. Milet, *Os Quebra Kilos*, pp. 4-7. This assumption biases the calculation against high profits, for one normally expects decreasing unit costs with increasing scale, within limits.

SOURCES: SAAP, *Relatório . . . de 4 de julho de 1878 . . . Ignacio de Barros Barreto e a Acta*, p. 39. *Idem., Trabalhos do Congresso Agrícola*, p. 228. *Relatório da Direcção da Associação Commercial Beneficente de Pernambuco apresentado à Assembléa Geral da mesma em 30 de agôsto de 1882*, Table. Emilio Dolé, "Engenhos Centraes," *Diário de Pernambuco*, September 27, 1890. For similar profit rates at the Usinas Pinto and Colonia Isabel, see Innocencio M. Araujo Alves, "Agricultura. Memorial Industria Saccharina," *ibid.*, November 18, 1891.

Before concluding this analysis of barriers to modernization, we must mention one factor which did not operate as expected. The government policy of low tariffs during the period of the trade treaties of 1810 and 1827 has been held partly responsible for the absence of a heavy machinery industry in Brazil in the first half of the nineteenth century.[26] In fact, the first steam engine in South America was built in the English-owned Aurora Foundry in Recife in 1829, and the foundry placed a similar motor in a Jaboatão engenho in 1836. This infant industry survived the virtually unimpeded importation of British machinery and received mild encouragement from the Alves Branco Tariff of 1844, which imposed duties of 30 to 35 percent on imported agricultural machinery, while permitting free entry of crude iron.[27] But despite the existence of this industry, and the low tariffs, prior to 1844, relatively few senhores de engenho modernized. When the government reduced tariffs on imported machinery in the late

---

drove them into debt, see Sr. Nabor, *Annaes da Assembléa Legislativa Provincial*, April 9, 1864, p. 445; and CP, "Os novos apparelhos de fabricar assucar," *Diário de Pernambuco*, September 19, 1881. Luis Amaral recognized that cost discouraged innovation. *História Geral da Agricultura Brasileira*, v. I, p. 358. Alice P. Canabrava, while stressing the abundance of cheap land and labor, blames the technological backwardness of the colonial period on successive market crises, an explanation which can be extended to the nineteenth century. "A Grande Propriedade Rural," Buarque de Holanda, *História Geral da Civilização Brasileira*, tomo I, A *Época Colonial*, v. 2. *Administração, Economia, Sociedade*, (São Paulo, 1960), p. 216.

26. Werner Baer, *Siderurgia e Desenvolvimento Brasileiro*, translated by Wando Pereira Borges (Rio de Janeiro, 1970, first published as *The Development of the Brazilian Steel Industry*, 1970), p. 78. Baer also mentions the absence of skilled workers. Palladius, "Os novos apparelhos de fabricar assucar," *Jornal do Recife*, August 19, 1881. CP, "Os novos apparelhos de fabricar assucar," *Diário de Pernambuco*, September 19, 1881. "Parte Industrial e Agricola, I," *ibid.*, August 22, 1893.

27. The first imported steam engine in Brazil was installed in Bahia in 1815. Deerr, *The History of Sugar*, v. II, p. 552. Compare *ibid.*, v. I, p.11,where Deerr gives the year 1813. Pereira da Costa, "Origens Históricas," p. xxvi. Flávio Guerra, *Idos do Velho Açúcar* (Recife, 1965), pp. 134-135. Vilela Luz, *A Luta pela Industrialização*, p. 22. This protection in 1844 did not function very effectively, for Brazilian imports of English sugar-mill machinery reportedly doubled between 1845 and 1847. William Law Mathieson, *Great Britain and the Slave Trade, 1839-1865* (London, 1929), p. 29.

1850's, as an anti-inflationary measure, and raised duties on raw materials to 5 to 30 percent, the local machinery manufacturers were forced to convert to simple repair shops. Exactly in this period, however, many mill owners began installing foreign-made steam engines. Rather than inhibiting industrial development, in other words, low tariffs seem to have encouraged it by facilitating imports. Indeed, Pernambuco's failure to import all the latest inventions in sugar manufacture in the 1840's and 1850's might be ascribed to the prevailing high tariffs.[28]

The relatively low costs of land and labor, low profit rates, the faltering world sugar market, and high tariffs all militated against capital-intensive innovation. Despite these obstacles, however, after 1870 a small group of mill owners began modernizing their mills with their own resources. Members of the sugar oligarchy, with many mills and sizeable incomes, could best afford the risk of modernizing. Francisco do Rêgo Barros de Lacerda started re-equipping his Engenho São Francisco da Várzea in 1873, and by 1874 he had spent a total of 33 contos on material plus 17 contos for transport and installation. The Barão de Muribeca, Rêgo Barros' uncle, began modernizing his Engenho São João da Várzea in 1875. He was the first planter to adopt vacuum pans and the multiple-effect. He paid 40 contos for the machinery, including the first vacuum crystalizing boiler in the province, plus 10 contos for freight and installation. Both mills in Várzea, a Recife suburb, enjoyed low transport costs.

Two senhores de engenho in the county of Escada kept pace with the enterprising mill owners in Várzea. In 1877 Antônio Marques de Holanda Cavalcanti refurbished his Engenho Mameluco, raising its capacity to 600 tons of sugar by 1881, and Francisco do Rêgo Barros de Lacerda declared "the Mameluco mill is uncontestably the best-equipped in the province." Antônio Marques' brother-in-law, Marcionilo da Silveira Lins, also

28. Vilela Luz, *A Luta pela Industrialização*, p. 24, 36. SAAP, Livro de Atas, no. 2, August 18, 1875. Herdeiros Bowman, "Aos Senhores de Engenhos," *Jornal do Recife*, March 25, 1880. Muitos operarios, "Os Srs. Bowman e a nova tarifa," *ibid.*, March 30, 1880. SAAP, *Acta . . . 28 de setembro de 1882 e Relatório*, pp. 15-16. For a list of technical improvements in the Cuban sugar industry between 1830 and 1860, see de la Sagra, *Cuba, 1860*, pp. 86-105.

imported better machinery, which enabled him to produce 800 tons of sugar per harvest in 1881. Antônio Marques' brother, Henrique Marques de Holanda Cavalcanti, Barão de Suassuna, took the lead in innovations: in the early twentieth century he introduced gasoline-powered tractors for cultivating and grading, installed steam and gasoline-powered pumps for irrigating, and donated lands to establish an experimental farm.[29]

The risks and difficulties of modernization were such that most modern mills in the nineteenth century enjoyed official subsidies. Before 1870, the provincial government had made a few sporadic attempts to stimulate the adoption of new technology. In 1843 and 1844 the provincial president contracted French engineers to introduce new clarifying agents, steam heating, and vacuum boilers. In 1847, Provincial Law 142 authorized the purchase of a new type of kettle from the French inventors Derosne and Cail. In 1851, the Pernambuco provincial president conceded a fifteen-year monopoly to Alfred and Edward De Mornay, who had patented a four-roller horizontal mill, which in conjunction with boilers and a centrifuge could increase productivity by 50 percent. In 1854 the provincial government imported and distributed new cane varieties, and granted twenty contos to the Monteiro sugar refinery of Rêgo and Barreto in Recife, to permit reconstruction after flood damages. This refinery used steam-heated vacuum boilers and a centrifuge, and the owners hoped to make it a training center.[30]

29. Francisco do Rego Barros de Lacerda, "Ao Sr. Palladius," *Diário de Pernambuco*, September 14, 1881. Um visitante, "Agricultura," *ibid.*, March 31, 1881. "Industria Agricola Assucareira em Pernambuco," *ibid.*, July 21-22, 1881. Milet, *Os Quebra Kilos*, pp. 105-107. Smith, *Brazil: The Amazons and the Coast*, pp. 443-447, describes a visit to the Engenho São Francisco da Várzea. "Melhoramento Agrícola", *Diário de Pernambuco*, August 24, 1877. SAAP, *Relatório Annual apresentado na sessão de 4 de julho de 1878 da Assembléa Geral pelo Gerente Ignacio de Barros Barreto e a Acta da mesma sessão* (Pernambuco, 1878), p. 39. *Idem.*, *Trabalhos do Congresso Agrícola do Recife em Outubro de 1878* (Recife, 1879), p. 228. "SAA," *Diário de Pernambuco*, February 6, 1894. Peres and Peres, *A Industria Assucareira em Pernambuco*, pp. 169, 172.

30. Pereira da Costa, "Origens Históricas," pp. xxvi-xvii, xlvi. "Interior," *O Progresso* (Recife), Tomo II (1847), p. 445. *Relatório com que fez entrega d'administração provincial o Exm. Sr. Victor de Oliveira, ao Exm. Sr. Francisco An-*

After 1870 the principal efforts by both local and national governments to promote modernization took the form of subsidies to builders of central mills and usinas. These subsidies transformed the Pernambuco sugar industry, and are described in Chapters Four and Five.

Transportation

The state of transportation technology directly affected the senhor de engenho because he bore the cost of transporting his sugar from mill to market. That cost fell after 1870 as a result of railroad expansion, but this expansion occurred slowly, on a limited scale, and at higher costs than anticipated. By the turn of the century, moreover, there were still barriers to the smooth flow of goods. Sugar could reach Recife by water and by land. Plantations near the ocean or along navigable rivers usually rented cargo space on sailing *barcaças* of 25 to 50 tons capacity, owned by Recife entrepreneurs. The sailing *jangada* represented an alternate form of water transport, but these rafts were smaller, slower, of lower freeboard, and "exposed to a thousand accidents."[31]

Sugar also traveled by land; as the number of engenhos increased in the western sugar zone and other areas without access to navigable portions of rivers or the ocean, land transport became increasingly important. In the early nineteenth century the main land conveyor was the wooden oxcart laden with two

tônio Ribeiro em 9 de março de 1852, p. 27. Deerr, A History of Sugar, v. II, pp. 545-546. Relatório que à Assembléa Legislativa Provincial de Pernambuco apresentou no dia da abertura da sessão ordinária de 1855 o Exm. Sr. Conselheiro Dr. José Bento da Cunha e Figueiredo, Presidente da mesma Província, pp. 1-7. Diégues Júnior, Populaçao e Açúcar, pp. 142-143.

31. José Alípio Goulart, Transportes nos Engenhos de Açúcar (Rio de Janeiro, 1959), pp. 41-44. Report by Bonham, April 30, 1881, Parliamentary Papers, 1881, HCC, v. XCI, AP, v. XXIII, p. 96. Joseph M. Stryker to William Hunter, Pernambuco, June 10, 1878, U.S. National Archives, Despatches from U.S. Consuls, v. 10, estimates barcaças' capacity at 50 to 200 tons. For a list of ten ports along Pernambuco's coast where barcaças loaded sugar freight, see Peres and Peres, A Industria Assucareira em Pernambuco, p. 202. Tollenare, Notas Dominicais, p. 69. Goulart, Transportes nos Engenhos, pp. 37-38. Recife and San Francisco Railway Company (Limited), Report of the Proceedings at the Thirty-First Half-Yearly Ordinary General Meeting of the Shareholders . . . April 18, 1871, p. 4.

chests weighing between one-half and three-quarters of a ton, each drawn by six to twelve oxen. Around mid-century, horse and mule caravans, each animal with a sixty to eighty kilogram sack slung on either flank, replaced the slow heavy oxcarts. The cotton sacks were at first imported, but by 1881 they were bought from local textile mills. In addition to freight charges, the muleteers also asked for a cask of molasses or a commission if they negotiated the sugar sale. Mules were preferred to oxcarts or sea-going transport. The mules moved faster than oxen, and avoided "special breakage" and wetting, thereby fetching prices 8 percent higher than barcaça-carried sugar.[32]

Public roads allowed, but hardly facilitated, overland transport. By the late 1870's these roads extended a total of 384 kilometers from Recife north, west, and south out across the sugar zone. Private roads supplemented this system: planters in Escada and Jaboatão built several kilometers to connect their five sugar mills with railroad stations in the 1870's. But road quality reduced utility and kept pack animal freight charges high. The road between Escada and Vitória, for example, "one of the most traveled in the province," was in such deplorable condition "that it tortures the heart. There the poor muleteers struggle with the greatest difficulties to force their animals to climb or descend the steep clay hills, all cut by deep ruts made by rainfall, and later they get caught in the flat lowlands with their fearful mudpatches and horrible bumps."[33]

32. Goulart, *Transportes nos Engenhos*, p. 80. Tollenare, *Notas Dominicais*, p. 68. Peres and Peres, *A Industria Assucareira em Pernambuco*, p. 27. Azevedo, *Canaviais e Engenhos*, p. 218. Report by Bonham, *Parliamentary Papers*, 1881, HCC, v. XCI, *AP*, v. XXIII, pp. 96, 114. *Falla com que o Exm. Sr. Commendador Henrique Pereira de Lucena abrio a sessão da Assembléa Legislativa Provincial de Pernambuco em 1º de março de 1874*, p. 60. Milet, *A Lavoura de Canna de Assucar*, pp. 37-38. Report by Hunt, August 18, 1864, *Parliamentary Papers*, 1865, HCC, v. LIII, *AP*, v. XXIV, p. 365.

33. Ceresiades, "A agricultura em Pernambuco, VII, IX," *Diário de Pernambuco*, July 2, 5, 1878. Many roads were built during the severe drought of 1877-79 by drought refugees *(retirantes)* hired by the province in make-work projects. Recife and San Francisco Railway Company (Limited), *Report of the Proceedings at the Fiftieth Half-Yearly Ordinary General Meeting of the Shareholders . . . October 12, 1880.* p. 2. For similarly critical descriptions of the roads, see Koster, *Travels in Brazil*, v. I, p. 66; Charles Blachford Mansfield,

To build railroads in Brazil, as to build central mills and usin-as, government subsidies were necessary. While European investors had introduced railroads in the Western Hemisphere colonies of Cuba in 1837 and Jamaica in 1845, independent Brazil had to wait until the 1850's. The first railroad, operated in 1854 in the province of Rio de Janeiro, was the government-owned Estrada de Ferro de Petropolis. The second Brazilian railroad was built in Pernambuco when the Imperial and provincial govern-ments guaranteed an English company, the Recife and San Fran-cisco Railway Company, Limited, 7 percent on its capital and a ninety-year monopoly to build a line from Recife southwest to the junction of the Una and Pirangí rivers. Construction began in 1855, and the line reached Palmares at kilometer 125 in 1862. The provincial government extended this line, under the name Estrada de Ferro do Sul de Pernambuco, 141 kilometers to Gar-anhuns in 1887. A branch line from Glicerio south into Alagoas was completed in 1894, and a group of planters built 47 kilome-ters of track from Ribeirão northwest to Bom Destino. This Recife and San Francisco Railway system depended principally upon sugar freight: "the sugar crop is the support of both coaching and goods traffic" declared the company directors. By the early 1900's the line was carrying up to one-half the state's sugar production (Table 13).[34]

Paraguay, Brasil, and the River Plate, Letters written in 1852-1853 (Cam-bridge, Eng., 1856), pp. 48-50; and Ulick Ralph Burke and Robert Staples, Jr., Business and Pleasure in Brazil (London, 1884), pp. 115, 119. Gilberto Freyre, Um Engenheiro Francês no Brasil (Rio de Janeiro, 1940), pp. 172-192, gives a summary of road conditions around the mid-nineteenth century. Célia Freire Aquino Fonseca, "Rotas, Portos, Comércio e a Formação do Complexo Açu-careiro em Pernambuco," Anais do V Simpósio Nacional dos Professores Universitários de História. Portos, Rotas e Comércio 2 vols. (São Paulo, 1971), v. 1, p. 359, points out that the poor road conditions inhibited direct sugar shipments to Recife before the advent of the railroad.

    34. Ademar Benévolo, Introdução à História Ferroviária do Brasil. Estu-do social, político e histórico (Recife, 1953), pp. 70, 80, 280. Estevão Pinto, His-tória de Uma Estrada-de-Ferro do Nordeste (Rio de Janeiro, 1949), pp. 57-61. Pinto gives dates for the inauguration of new stations, pp. 247-252. In view of this railroad's early construction date, it seems unlikely that the imperial guarantees represented the government's attempt to ease the inflation of the later 1850's and early 1860's, as suggested by Vilela Luz, A Luta pela Industrialização, p. 25. More likely, the imperial government encouraged the railroad at this time to

In the 1880's, the railroad network stretched to the west and northwest. Another English company, the Great Western of Brazil Railway Company, Limited, received the 7 percent imperial guarantee and laid two tracks into the province's northwestern sugar districts. The main line ran from the capital west 84 kilometers to Limoeiro, and a secondary line branched from Carpina north 58 kilometers to Timbaúba. In 1901 this line was connected to the railroad in southern Paraíba. This Great Western system carried between one-fifth and one-third of the sugar shipped to Recife.[35]

To connect the western sugar mills and the cotton plantations with Recife, the Imperial government acceded to requests by senhores de engenho and provincial authorities and in 1881 began building the Estrada de Ferro Central de Pernambuco. By 1887 this government-owned railroad ran 180 kilometers west from Recife through the sugar zone to Antônio Olinto, but it carried only 5 percent of the crop.[36]

As a result of railroad growth, trains were soon carrying three-quarters of all sugar shipped to Recife and they had virtually eliminated the pack animal caravans, which moved less than 5 percent of the crop after 1885. The barcaças remained the principal alternate carrier, but their share of sugar freight declined from over one-third to under one-fifth by 1910.

The sugar planters welcomed the extension of railroads into

cushion the negative effects of stopping slave imports in 1850. The initial guaranteed capital was £1.2 million, not £900,000 as noted by Alfred Tischendorf in "The Recife and San Francisco Pernambuco Railway Company, 1854-1860," *Inter-American Economic Affairs*, v. XIII, pp. 4 (Spring 1960), pp. 87-94. See Cowper to Clarendon, April 15, 1856, *Parliamentary Papers*, 1857, HCC, v. XLIV, AP, v. XX, p. 233. Walle, *Au Brésil du Rio São Francisco a l'Amazone*, p. 152. "Empresa da Estrada de Ferro de Ribeirão à Bonito em 10 de Maio de 1893," *Diário de Pernambuco*, June 9, 1893. Recife and San Francisco Railway Company (Limited), *Report of the Proceedings at the Forty-First Half-yearly Ordinary General Meeting of the Shareholders . . . December 31, 1875*, p. 3.

35. Benévolo, *Introdução à História Ferroviária*, p. 80. Pinto, *História de Uma Estrada-de-Ferro*, pp. 83-84, 91-92. That the Recife-San Francisco Railway carried more sugar contradicts Richard Graham's contention that the area traversed by the Great Western "northwest of Recife was richer than that of the Recife and São Francisco" (sic, the company's name was anglicized). Graham, *Britain and the Onset of Modernization*, p. 70.

36. Pinto, *História de Uma Estrada-de-Ferro*, pp. 103-104.

TABLE 13

SUGAR TRANSPORT

Conveyor's Share of Total Sugar Shipments to Recife (percent)

Year[a]	Barcaças[a]	Steamers[b]	Animals	Recife-San Francisco RR	Great Western RR	Estrada de Ferro Central	Olinda, Caxangá RR
1883	36.4	3.7	8.6	36.1	18.6		0.3
1885	40.6		6.1	42.9	7.7	2.7	
1886	40.3		4.6	37.7	13.4	4.0	
1887[c]	36.4		3.2	39.1	16.7	4.6	
1888	36.9		4.1	37.2	16.8	5.0	
1889	38.2		4.6	41.3	12.1	3.8	
1890	38.2		3.9	36.4	16.8	4.1	
1891	38.3		3.9	37.9	15.1	4.8	
1892	37.7		3.6	38.0	16.1	4.6	
1893	33.6		3.1	36.9	21.9	4.5	
1894	32.8		2.1	36.6	22.8	5.7	
1895	32.6	0.03	1.8	39.2	20.9	5.5	
1896	20.7		1.8	42.9	29.2	5.4	

1897	23.5	0.3	2.0	43.4	26.1	5.0	
1898	19.9	0.1	2.4	46.4	24.3	4.0	0.6
1899	15.1	0.6	1.4	41.4	36.6	4.6	0.8
1900	24.8	0.04	2.7	48.9	18.8	4.2	0.5
1901	24.8	3.1	1.8	38.6	24.6	4.7	0.8
1902	27.1	0.9	2.3	46.6	17.8	4.4	0.9
1903	18.1		2.0	44.2	32.2	3.0	0.5
1904	25.8	0.6	1.8	49.0	16.3	4.6	2.0
1905	17.0		1.2	44.3	32.8	3.6	1.2
1906	20.0		1.2	49.0	25.9	3.3	0.6
1907	14.7		1.7	64.4	15.1	3.9	0.9
1908	20.2		0.5	50.6	24.0	4.3	0.8
1909	14.6		0.04	50.7	32.0	2.4	

[a] Data for 1883-1897 is for the harvest beginning in the year indicated. Data for 1898-1909 represents average shares of shipments for October, January, and July, which were respectively the beginning, middle, and aftermath of the harvest.

[b] Steamers carried sugar from other provinces, and frequently were excluded from reports.

[c] In 1887 and 1888, we include Goiana Canal traffic as barcaças.

SOURCES: The Great Western of Brazil Railway Company, Limited, *Report of the Directors and Statement of Accounts to 30th June 1883 to be Submitted to the Shareholders at the Annual General Meeting to be Held on the 31st October, 1883*, through idem., *Report of the Directors and Statement of Accounts for the Year Ended 31st December, 1909, to be submitted to the Shareholders at the Annual General Meeting*.

the *zona da mata*, the principal growing region. They expected
the railroad would charge lower freight rates than other conveyors
and would provide better service. They also hoped that the railroad would put the 20,000 muleteers into the ranks of the rural
unemployed, thereby forcing down wages. Others anticipated
that the railroads' westward penetration of the *agreste* and *sertão*
regions would bring food crops into Recife markets. Before the
railroads' arrival, corn and beans had been imported from Portugal and Italy, and beef from Rio Grande do Sul, Argentina, and
Uruguay. The merchants' ACA hoped that the railroad would
allow interior farmers to sell their crops in Recife for prices below
those of imported goods, and thereby raise the standard of living,
reduce the cost of living, and put more downward pressure on
wages.[37]

The railroads fulfilled some of these expectations. Railroad
freight rates undercut mule train rates by about one-half. But
many still thought the rates high. Resentments concentrated on
the two English companies. Planters and others repeatedly criticized the Recife and San Francisco Railway for overcharging both
customers and the imperial government by maintaining high
freight rates while at the same time receiving sizeable official
subsidies. Between 1858 when the line began, and 1883, the
Company had earned profits totaling 7,530 contos and had
received imperial subsidies totaling 14,576 contos.[38]

The Recife and San Francisco Railway experienced financial
difficulties for several reasons. One Brazilian engineer implied
deliberate fraud, and both the U.S. consul and a major Brazilian
stockholder agreed that the railroad's route "serves the purposes

37. SAAP, *Trabalhos do Congresso Agrícola*, pp. 156-157, 174-175, 277:
Luiz de Carvalho Paes de Andrade, *Questões Econômicas em Relação à Província de Pernambuco* (Recife, 1864), p. 55.

38. "Sobre o ínfimo preço do assucar," *Diário de Pernambuco*, September
26, 1886. "Tarefas das vias ferreas. Parecer apresentada à Assembleia Geral da
Sociedade Auxiliadora da Agricultura de Pernambuco em 25 de março de 1882
(*sic*), pelo presidente interino da sessão de economia social e rural da mesma Sociedade engenheiro Henrique Augusto Milet," *ibid.*, April 14-16, 1887. "Parte
Official. Ministério da Agricultura," *ibid.*, June 28, 1888. Cyro Diocleciano Ribeiro Pessôa Junior, *Estudo Descriptivo das Estradas de Ferro do Brazil Precedido da Respectiva Legislação* (Rio de Janeiro, 1886), pp. 100-101.

of a few individuals" but not "the greater part of the planting population." Others blamed the wide 1.6 meter gauge, which cost more to build and maintain and prevented speeds above 40 kilometers per hour. Moreover, since all the other Pernambuco railroads, including the Sul de Pernambuco which extended the Recife and San Francisco to Garanhuns, ran on standard 1 meter gauge, transshipment between the Recife and San Francisco and other lines required costly reloading, and the former could not share rolling stock with the others.[39]

Poor service drew as many complaints as high freight rates. One Escada senhor de engenho wrote the imperial government to protest "the deplorable state" of the Recife and San Francisco Railway; he complained of "delays of a week or two in the interior stations . . . where raw and purged sugar are stored haphazardly in the few warehouses, the baggage and the passenger rooms, and even piled on the station platform edges, exposed to the weather and to theft, waiting for shipment."[40]

Because the imperial government guaranteed a profit to the English railroad companies, it could impose certain requirements

39. Ribeiro Pessôa Junior, *Estudo Descriptivo*, p. 119. Thomas Adamson Jr. to William H. Seward, Pernambuco, November 14, 1864, U.S. National Archives, Despatches from U.S. Consuls, v. 7. Visconde de Mauá, *Autobiografia*, cited in Graham, *Britain and the Onset of Modernization*, p. 69. Graham concluded that "traffic was not heavy enough to make the enterprise pay." Alfred Marc, *Le Brésil. Excursion à Travers des 20 Provinces*, 2 vols. (Paris, 1890), v. I, p. 251. Coelho Cintra, "Congresso Nacional—Discurso Pronunciado na Sessão de 22 de Novembro de 1895," *Diário de Pernambuco*, March 5, 1896. A.F. Howard, "Report for the Year 1896 on the Trade, etc., of the Consular District of Pernambuco," *Paliamentary Papers*, 1897, HCC, v. LXXIX, *AP*, v. 28, p. 15. The railroad itself confessed that it lacked adequate rolling stock before 1881. Recife and San Francisco Railway (Limited), *Report of the Proceedings at the Fifty-First Half-Yearly Ordinary General Meeting of the Shareholders . . . April 5, 1881*, p. 1. According to foreign visitors, the Great Western Railway rates in the 1880's were also considered exorbitant, although the company escaped charges of deliberate fraud. James W. Wells, *Exploring and Travelling Three Thousand Miles Through Brazil From Rio de Janeiro to Maranhão*, 2 volumes (London, 1886), v. II, p. 338. Marc, *Le Brésil*, v. i, p. 250.

40. SAAP, Livro de Atas no. 2, February 3, 1881. Mixing different sugar types promoted product deterioration. This complaint echoed an earlier protest by a planter who had to leave his slaves to sleep on top of the sugar sacks to prevent theft. Antonio Venancio Cavalcante de Albuquerque, "Agricultura ou a questão da actualidade," *Diário de Pernambuco*, April 7, 1877.

in response to planter complaints. In 1882 the Agriculture Minister ordered the Recife and San Francisco to increase its rolling stock out of its reserve fund—that is, without any increase in capital guaranteed. The Rio de Janeiro authorities also set freight rates in 1857, 1863, and 1888. But these adjustments proved only temporary solutions. In 1900 the federal government finally bought out the Recife and San Francisco Railway Company.[41]

The government's principal scheme for improving the Pernambuco railroads entailed leasing the Recife and San Francisco and the Sul de Pernambuco in 1901, the Central de Pernambuco in 1904, and the Ribeirão-Bonito in 1905, to the Great Western, the only profitable English railroad in the northeast, and conceding import duty and tax exemptions. As a result of this consolidation, service reportedly improved and the Great Western linked together all the Pernambuco lines.[42]

Once the sugar had reached Recife, there still remained two short but troublesome transports before it was on the boat for Europe, the U.S., or southern Brazil. The Recife and San Francisco line ended at Cinco Pontas, an old fort about one kilometer from the warehouses and ocean shipping docks. In the early 1860's, the railroad had attempted transshipping by water through the harbor, "but met with so much opposition that they were obliged to abandon it." Later they tried oxcarts, but these carried only eight sacks each and were "limited in number and uncertain in attendance." In the later 1870's, the railway con-

41. SAAP, *Acta . . . 28 de setembro de 1882 e Relatório*, p. 26, Recife and San Francisco Railway Company (Limited), *Report of the Proceedings . . . June 30, 1882*, p. 3. Milet, *A Lavoura da Canna de Assucar*, pp. 29-30. "Redução de Taxa," *Diário de Pernambuco*, June 19, 1887. "50 % de abate no transporte de cannas," *ibid.*, August 3, 1887. "Parte Official. Ministério da Agricultura," *ibid.*, June 28, 1888. Consul Howard, "Report on the Trade and Commerce of the Consular District of Pernambuco for the years 1899-1900," *Parliamentary Papers*, 1901, HCC, v. LXXXI, *AP*, v. 45, p. 8.

42. Consul Staniforth, "Report on the Trade and Commerce of Pernambuco for the Year 1905," *ibid.*, 1906, HCC, v. CXXIII, *AP*, v. 59, pp. 5-6. Consul George A. Chamberlain, "Pernambuco," U.S. Department of State, *Reports from the Consuls of the United States on the Commerce, Manufactures, etc., of Their Consular Districts*, 1908, v. II, p. 327. Consul Pearson, "Report on the Trade of the Consular District of Pernambuco for the Year 1911," *Parliamentary Papers*, 1912-13, HCC, v. 94, *AP*, v. 46, p. 7.

tracted with a Brazilian tramway company, the Locomotora, which used burros to pull open cars over iron rails. But the carters and the Recife sugar factors who held shares in carting companies fought the Locomotora, often physically blocking the tramway's movement, and the company failed. By the early 1880's, the railway returned to supervising transshipment by wagon, but the sugar remained "liable to new damage": the open wagons let rain in, ripped sacks, and spilled sugar. The Central de Pernambuco line presumably faced similar problems, for its terminal was located even farther from the harbor docks. Only the original Great Western terminal near the north end of the island of Recife was close to the shipping docks.[43]

At the docks, the sugar exporter faced the port's principal geographical disadvantage. While the long reef sheltering the harbor provided an excellent protected anchorage, the bar at the harbor entrance left less than four meters of water at low tide. This shallow entrance forced both passengers and freight aboard trans-Atlantic sailing ships to disembark in smaller canoes, lighters, and even jangadas; virtually every nineteenth-century visitor remarked on this inconvenient adventure. The arrival of steamships accelerated the pace of international commerce, but these larger and deeper-draft ships could not cross the bar even at high tide, and remained totally dependent upon the lighters. Even smaller steamers entering the harbor had to use lighters, for at dockside the water was shallow. For over eighty years Pernambuco governments commissioned studies of port improvements, but little was done, despite repeated protests by the ACBP, apart from slow dredging which deepened the inside channel.[44]

43. Recife and San Francisco Railway Company (Limited), *Report of the Proceedings at the Fifteenth Half-Yearly Ordinary General Meeting of the Shareholders . . . April 1863*, p. 5. Mário Sette, *Arruar, História Pitoresca do Recife Antigo*, 2nd ed. (Rio de Janeiro, n.d., first published 1948), pp. 94-96. Raffard, *O Centro da Indústria*, p. 119.

44. Hadfield, *Brazil, The River Plate and the Falkland Islands*, p. 102. Joseph M. Stryker to William Hunter, Pernambuco, November 10, 1877, April 12, 1878, U.S. National Archives, Despatches from U.S. Consuls, v. 10. Frank Bennett, *Forty Years in Brazil* (London, 1914), p. 3. Consul Howard, "Report on

While transportation grew cheaper and the daily wage level
for unskilled rural labor fell after 1870, possibly because of the
increased numbers of unemployed muleteers, the cost of living
still rose. Little incentive existed to extend the railroads into the
sertão. Several writers criticized the railroads as "defective due
to the tracks' proximity to the coastline," but both the provincial
government and the private companies preferred to operate railroads which would "take better care of the interests of agriculture
by serving populated and productive regions." The 1872 and
1890 censuses showed that the sertão contained less than one-
fifth the population, whereas the zona da mata alone held over
one-half. Nor did the sertão produce a major export to provide
freight revenues.[45]

The modernization of sugar technology in Pernambuco only
began in earnest in the 1870's. Productivity estimates in terms of
both sugar worker and total per capita sugar output illustrate well
the rate of technical progress. Productivity remained fairly constant during the first forty years of the nineteenth century. After
1840, productivity rose gradually and accelerated after 1870.
Only when the export market collapsed did productivity level off
(Table 14).

---

the Trade and Commerce of the Consular District of Pernambuco for the Years
1899-1900," *Parliamentary Papers,* 1901, HCC, v. LXXXI, *AP,* v. 45, p. 7. Act-
ing Consul Williams, "Report on the Trade and Commerce of the Consular District of Pernambuco for the Year 1901," *ibid.,* 1902, HCC, v. CV, *AP,* v. 51, p.
8. For a bibliography of port improvement proposals, see F.A. Pereira da Costa,
Anais Pernambucanos, 10 volumes (Recife, 1949-66), v. X, pp. 481-487. Gilberto Freyre recalls that the provincial government had sought foreign engineers to
improve the port since 1825. Freyre, *Um Engenheiro Francês no Brasil,* pp.98-
99.

45. Ceresiades, "A agricultura em Pernambuco, VII, VIII" *Diário de Pernambuco,* July 2-4, 1878; SAAP, *Trabalhos do Congresso Agrícola,* p. 139.
Milet, *Os Quebra Kilos,* p. 69. Pinto, *História de Uma Estrada-de-Ferro,* p.
104. *Recenseamento da População do Império do Brasil a que se procedeu no
Dia 1 de Agôsto de 1872, Quadros Estatísticos,* 23 vols. (Rio de Janeiro, 1873-
1876), XIII, 214, Directoria Geral de Estatística, *Sexo, raça e estado civil, nacionalidade, filação culto e analphabetismo da população recenseada em 31 de
dezembro de 1890* (Rio de Janeiro, 1898), pp. 94-99. Definitions of the three
areas are found in *Anuário Estatístico de Pernambuco,* Ano XIX (Recife, 1964),
pp. 35-37.

TABLE 14

PRODUCTIVITY IN SUGAR IN PERNAMBUCO

Year	Kilograms/Plantation Worker [a]	Kilograms Per Capita [b]
1700-1800	825	
1810		28.6
1814		19.5[c]
1815	682	16.9
1817	682	
1819		24.8
1823		25.4
1832		24.9
1839		42.7
1842		50.2
1845	1,350	
1847	1,068	
1854	1,661	
1855		91.3
1857	2,810	
1872		116.7
1876	3,018	
1890		151.6
1900		114.0
1907[d]	18,058	

[a] Where a range of estimates was given, we have listed the arithmetic average.

[b] Sugar production taken from five-year averages.

[c] Declines in productivity may be attributed to underenumeration in the earlier censuses.

[d] This figure is only for 41 usinas, where heavy capital investments permitted such high productivity.

SOURCES: *1700-1800*: Simonsen, *História Econômica do Brasil*, pp. 133-134. *1815*: Koster, *Travels in Brazil*, v. II, p. 138. *1817*: Tollenare, *Notas Dominicais*, p. 74. *1845*: D. P. Kidder, *Sketches of Residence and Travels in Brazil*, 2 volumes (Philadelphia, 1845), v. 2, p. 133. *1847*: Report by de Mornay, 1847, in Deerr, *The History of Sugar*, v. II, p. 358. *1854: Relatório . . . 1854 . . . José Bento da Cunha e Figueiredo*, table. *1857: Relatório . . . 1857 . . . Sérgio Teixeira de Macedo*, pp. 75-77. *1876*: Milet, *Os Quebra Kilos*, pp. 4-6. *1907*: "Quadro Estatistico da Indústria Assucareira no Estado do Pernambuco," *Boletim USAP*, Anno I, v. I, no. 5 (May 1907), pp. 386-387.

All data on kilograms per capita are from our Tables 5, 22, and 33.

At mid-nineteenth century the technological level in the Pernambuco sugar industry lagged considerably behind that of other cane sugar producing areas. Successive innovations, especially after 1870, reduced this lag, and by the early 1900's several dozen large modern usinas were operating (see Chapter Five).

Railroads carried cane from field to mill and sugar from mill to port, with considerable reductions in transport costs. The technological progress was most impressive in sugar manufacture and transportation, and least so in sugar cane agriculture. But at least the advances had occurred. It was not technological backwardness that caused the sugar industry's stagnation.

4
CAPITAL MOBILIZATION
THROUGH TRADITIONAL SOURCES

$\mathcal{A}$ll the gentlemen addressing the 1878 Recife Agricultural Congress agreed that the northeastern sugar economy needed more capital.[1] Capital could serve to introduce more modern technology in factory and field, to produce a greater volume of better quality sugar at a lower cost, and thereby, it was hoped, to recover export markets. But the agreement on the problem was not accompanied by unanimity with respect to its solution. Mortgage law reform, more banks, increased emissions or loans, and business association law reform all gained adherents and sometimes official support. The most effective instrument proved to be the official subsidy, whether indirectly through a bank or directly to the planter.

The Correspondente

Traditionally, the sugar planter raised capital from only one source, a Recife factor known as the *correspondente* or *comissário*. At the beginning of the planting season, the correspondente advanced cash to the planter, who paid up to 6 percent interest per month on short-term loans.[2] The correspondente

1. SAAP, *Trabalhos do Congresso Agrícola*, pp. 114, 144, 175, 219, 225, 279, 317, 441.
2. "The interest rate in some provinces ranges from 7 percent to 12 percent, and in others from 18 percent to 24 percent, and in still others from 48 percent to 72 percent" for periods up to one year. Commissões da Fazenda e Especial, *Parecer e Projecto sôbre a Creação de Bancos de Crédito Territorial e Fabricas Centraes de Assucar apresentados à Câmara dos Srs. Deputados na Sessão de 20 de Julho de 1875* (Rio de Janeiro, 1875), p. 36.

himself borrowed his money from the commercial banks, where
prime interest rates after 1850 usually fluctuated between .75
and 1 percent monthly (Table 15). Because he faced greater risks
than the correspondente, the planter could not borrow directly
from the bank and had to rely on the middleman. The correspondente also usually bought the planter's sugar and rum, or acted as
selling agent, in return for a 3 percent commission on the sale
price.

Occasionally the correspondente allowed the sugar planter

TABLE 15

PRIME INTEREST RATES, RECIFE

(in percent)

Year	Average Annual Rate	Year	Average Annual Rate
1835	18.00	1866	10.25
1836	18.00	1867	9.04
1837	16.00	1868	8.67
1838	15.13	1869	9.88
1839	16.13	1870	10.33
1840	14.38	1871	9.98
1841	13.50	1872	9.63
1857	9.43	1873	11.08
1858	11.00	1874	11.13
1859	11.96	1875	11.25
1860	14.63	1876	9.75
1861	13.04	1877	9.75
1862	14.42	1878	9.25
1863	10.31	1879	8.33
1864	8.96	1880	8.96
1865	10.42		

NOTE: The average annual rate was determined as the simple mean of monthly
quotations. If interest were compounded monthly, then the average annual rate would
be slightly higher.

SOURCES: Monthly quotations in *Diário de Pernambuco*. I am indebted to David
Denslow for permission to use this data.

reciprocal interest—that is, the planter would earn interest on the value of sugar and rum received by the correspondente for which the planter had not yet been paid. Finally the correspondente acted as the planter's purchasing agent in Recife, to order and ship food, clothing, machinery, tools, and other goods to the engenho, with a 3 percent commission on purchase prices for services rendered, plus 10 percent interest on the price if the correspondente paid for the goods with his own money. Gilberto Freyre has described the powerful correspondente:

In the eighteenth century and all through the nineteenth the power of the middleman, which had begun in the seventeenth century, grew. His status was dignified by his development into agent, commission merchant of sugar or coffee, and then banker. A city aristocrat, with a gold chain about his neck, silk hat, a tiled mansion, a luxurious carriage, eating imported delicacies, raisins, figs, prunes, drinking Port wine, his daughters ravishingly attired in dresses copied from the Parisian fashion books when they attended the *premières* of Italian divas at the opera house.[3]

In the later nineteenth century, the number of correspondentes varied with the sugar harvest: in leaner years, less than thirty correspondentes were listed in Recife almanacs, while during fatter years as many as one hundred were so listed. With 1,500 to 2,000 engenhos in the province, an average correspondente might service up to seventy accounts.[4]

During good years, planters could bear these middlemen's charges. During bad years, however, when planters could not pay their debts with cash, the correspondentes took slaves or mortgages in payment, or refused to furnish supplies, "with the result that mills shut down, consumption was limited, commerce atrophied, businesses failed, general malaise for all."[5]

3. Gilberto Freyre, *The Mansions and the Shanties (Sobrados e Mucambos), The Making of Modern Brazil*, translated by Harriet de Onís, (New York, 1963, first published 1936), p. 15.

4. José de Vasconcellos, *Almanack Administrativo, Mercantil e Industrial da Província de Pernambuco para o Anno de 1860; idem., ibid., 1861; 1862.* F.P. do Amaral, *Almanack Administrativo, Mercantil, Industrial e Agrícola da Província de Pernambuco para o anno 1868; idem., ibid., 1869, 1870, 1872, 1873, 1875, 1876, 1881,1884, 1885, 1886.*

5. SAAP, *Trabalhos do Congresso Agrícola*, p. 337.

Especially after 1870, as Pernambuco's sugar economy began feeling the pinch of foreign competition, debt-ridden planters frequently damned the correspondentes' lending practices, calling them "usurers," "sharpers," and "social harpies." Repeatedly the senhores de engenho protested against the "infernal usury" and "monstrous advances" which "in combination with the warehouser drain all the sweat from the poor planters." As one wrote: "Sugar is not profitable, but the warehouser and the correspondente accumulate colossal fortunes, and own palaces with very rich furniture and many household servants. They take vacations and enjoy the true life of nobles, whereas the poor planter who makes sugar barely survives, and even then only by virtue of a Franciscan thrift."[6]

Planters resented the middlemen's purchasing practices as well. The correspondentes sometimes paid prices below published market quotations. Moreover, since they could not raise the selling price of sugar in the international market, the middlemen passed part of sugar export taxes on to the producers, in the form of lower buying prices. When that export tax rate fell, middlemen's profits rose, unless competitive pressure obliged them to pass the additional money along to the producers in the form of higher prices. Thus, the merchants' ACBP and ACA, as well as the planters' SAAP, campaigned actively against the sugar export tax, and with some success.[7] Nevertheless, a few planters

6. Herculano Cavalcanti de Sá e Albuquerque, "Elemento servil," *Diário de Pernambuco*, September 22, 1871. "Breves considerações sôbre a agricultura do Brasil, XI," *ibid.*, May 13, 1876. Antonio Venancio Cav.te de Albuquerque, "Agricultura ou a questão da actualidade, III," *ibid.*, April 13, 1877. *Relatório da Direcção da Associação Commercial Beneficente de Pernambuco apresentado à Assembléa Geral da mesma em 6 de agosto de 1875, passim.*

7. Um victima das victimas, "Banco de Crédito Real," *Diário de Pernambuco*, May 28, 1887. For a reply, see Uma das victimas, "Banco de Crédito Real," *ibid.*, June 1, 15, 1887. The class organizations persuaded the provincial government to abolish the 4 percent export tax on sugar and cotton in 1875-76, and the imperial government reduced its own export tax on sugar from 9 percent to 7 percent for that harvest. The following year, however, the provincial export tax was reestablished, and subsequently raised as high as 3 percent. The imperial export tax remained between 5 and 7 percent until the end of the Empire, despite continued protests. SAAP, Livro de Atas no. 2, August 18 and 24, 1875; *Relatório . . . ACBP . . . 6 de agôsto de 1875*, p. 80ff.; *Projecto de Receita*

charged oligopsony, noting bitterly that "only the merchants make profits from the suppression of the export tax, since they impose the price on the producer with immunity.[8]

The planters were only partly justified in excoriating their middlemen for charging high interest. If one divides the years 1859-80 into three sub-periods, one finds wide fluctuations in the consumer price index (Table 26), including a swift fall from 1859 to 1864 and a slower but steady recovery from 1864 to 1878. The interest rate, on the other hand, experienced a very slow decline. The price of capital sometimes followed consumer price movements, which mitigated the planters' complaints.

If one measures usury by the profit rates of the borrowers, on the other hand, the planters had good reason to complain. Since some traditional engenhos returned net profits no greater than 9 percent, a mill owner with traditional technology might have to pay interest charges on his operating capital which exceeded his rate of return. He might be wiser, therefore, to forget sugar and become a correspondente; in fact, many wealthier planters did indirectly become creditors of their poorer colleagues.

The correspondentes could exact high interest for two reasons. First, loans to planters did entail higher risks:

The merchant, whose office and transactions are under his creditors' [the commercial banks] daily inspection and observation, guarantees punctuality of payment and swiftness of collection. But the planter, whose place of work is almost always distant, whose regularity of payment depends often on the contigencies and eventualities of a good or bad harvest, cannot offer as security either the pending harvest nor future harvests. The planter does not have, like the English farmer, books and a regular accounting system, and is not a strong attractor of capital.[9]

Provincial organizado por ordem do Excm. Desembargador José Manoel de Freitas, Digníssimo Presidente desta Província, pelo Administrador do Consula-do Provincial Bacharel Francisco Amynthas de Carvalho Moura (Pernambuco, 1884) p. 7.

8. "Breves considerações sobre a agricultura do Brasil, XIII," Diário de Pernambuco, May 24, 1876. The ACBP, not unexpectedly, disputed the notion that only middlemen benefited from reduction of the export tax. "Associações. ACB," Jornal do Recife, April 19, 1876.

9. Commissões da Fazenda e Especial, Parecer e Projecto, p. 43.

Personal Assets

Sugar planters usually had a poor cash position: their wills in the later nineteenth century infrequently bestowed cash upon the legatees (Table 16). Rather the planters held their wealth in land and associated productive factors; even luxury goods represented a very small share of total worldly goods.

TABLE 16

ASSETS OF SENHORES DE ENGENHO

Asset	*Frequency of Inclusion in Wills* [a] *(percent)*	*Average Proportion of Total Assets (percent)*
Engenhos, Properties, Land, Houses	100	65.7
Slaves	97 [b]	26.5
Livestock	86	5.5
Furniture, Household Goods	75	0.9
Cane, Sugar	68	7.5
Gold, Silver, Copper, Jewelry	49	2.2
Debts Held	25	6.8
Cash, Bonds	19	15.9
Wagons, Boats	15	3.6 [c]

[a] We analyzed 57 wills by senhores de engenho dated between 1859 and 1910.

[b] Thirty wills were dated prior to abolition.

[c] The total percentage of assets exceeds 100 percent because certain items, such as cash bonds, bulked disproportionately large in a few wills.

SOURCES: Inventories in wills archived in the Cartório Público de Ipojuca, Pernambuco.

Planters could raise capital by converting these assets into cash. But such liquidity in exchange for movable goods was generally inadequate for investment purposes, while real estate, until almost the end of the Empire, did not constitute good collateral. Direct sales of livestock and slaves brought cash, but individual livestock were worth relatively little, so a planter had virtually to empty his corrals and stables to raise even a little investment capital.[10] Slaves, on the other hand, definitely constituted a source of working capital for planters, who sold them at the rate of a few each year in the interprovincial slave trade after 1850 (see Chapter Seven). Neither livestock nor slave sales, however, resulted in amounts necessary to modernize traditional engenhos; moreover, liquidating these assets required replacing their productive function with machinery or free labor, which also usually required cash.

The abolition of the international slave trade in 1850 meant that capital formerly reserved for replenishing the labor force with fresh African imports could be put to other uses. Much of this capital in the center-south provinces entered coffee production, industries, and banks. In Pernambuco, some of the capital formerly employed in the international slave trade may have entered industry: a candle factory was inaugurated in 1860, an explosives plant in 1861, and nine better-quality cotton textile mills had appeared by 1866. The cotton boom of 1860-70 undoubtedly attracted such funds. Many of the four hundred new engenhos built in the later 1850's and the 1860's may have used slave trade capital. But the absence of modern machinery in Pernambuco mills before the 1870's indicates that slave trade capital did not pay for modernizing that sector.[11]

The occasional appearance of bank accounts and government bonds and debts held among planter's assets confirms that their initial reluctance to invest in modern machinery did not derive from simply a routinist mentality. One observer complained of "the egotism of many of our planters who, having capital, instead

10. In the 1850's, cattle were worth about 20$000 per head, and horses about 25$000 per head. By the end of the Empire, these prices had risen but few animals were valued above 100$000. Cartório Público de Ipojuca, Inventories in Ipojuca wills.

11. José Honório Rodrigues, "A Revolução Industrial Açucareira. Os Engenhos Centrais, I," *Brasil Açucareiro*, v. XXVII, no. 2 (February 1946), p. 81. Werneck Sodré, *História da Burguesia Brasileira*, pp. 109-110. Francisco Iglesias, "Vida política, 1848-1868," in Buarque de Holanda, *História Geral da*

of trying to organize a bank or a mutual aid society, prefer to keep their capital in the commercial banks, so that these can supply needy planters, through loans by the correspondentes who earn profits twice, thrice, and sometimes quadruple the interest paid by the banks to their depositors."[12]

The planters also invested in national and provincial or state bonds. Paying 6 to 7 percent annually, these bonds were transferable, easily convertible to cash, tax exempt, and virtually riskless. In 1878 Pernambuco residents reportedly held between 4,000 and 5,000 contos worth of such bonds. Only ten years later, the provincial assembly authorized the provincial president to borrow 7,500 contos to redeem outstanding 7 percent bonds. Some termed the imperial treasury "an omnipotent competitor" for capital, and labeled the government bonds "a gnawing cancer . . . a true whirlpool sucking in all the available capital in Brazil." The ACBP regretted "that in this province there is no tendency to employ this capital in enterprises of public utility, instead of discounting commercial paper and buying public bonds."[13]

The absence of large cash holdings among the planters' assets does not, of course, imply that the Pernambuco elite did not live well. If they had little cash to leave to their heirs, they certainly spent money on luxury items and services while alive. Wealthy senhores de engenho had often attracted attention by their lavish "Asiatic" spending. Elaborately furnished town and country homes, fast horses and fancy carriages, and, around the turn of the twentieth century, automobiles, characterized the "aristrocratic" life style of the richer planters.

Civilização Brasileira, tomo II, v. 3, *Reações e Transações*, pp. 35-36. Apollonio Peres and Manuel Machado Cavalcanti, *Industrias de Pernambuco* (Recife, 1935), pp. 77, 115, Raul de Góes, *Um Sueco Emigra para o Nordeste*, 2nd ed. (Rio de Janeiro, 1964), p. 44. A summary of the few light industries in Recife before 1850 appears in Pereira da Costa, *Anais Pernambucanos*, v. X, pp. 275-277. For new engenhos see our Table 21.

12. "O dr. José Antonio de Figueiredo ao público, XXVII," *A Provincia*, February 27, 1875.

13. SAAP, *Trabalhos do Congresso Agrícola*, pp. 19, 116, 123, 216. "Emprestimo externo de 8,600 contos," *Diário de Pernambuco*, November 15, 1888. *Relatório . . . ACBP . . . 30 de agôsto de 1882*, p. 16.

At the banquets of the wealthiest or most ostentatious planters—who since the sixteenth century had amazed Europeans by the lavishness of their entertainment—the wine flowed freely. Food was so abundant that it went to waste; at the end came the musical toasts. And wine everywhere, on the tablecloth, on the floor, a display of conspicuous waste. Old Major Santos Dias of Jundiá [an Escada sugar plantation], was one of the last of the planters who was renowned for his bountiful board . . . English lords who came out to Pernambuco to hunt cougars in the plantation forests were guests at Jundiá. . . .[14]

But luxury articles, at least those which were passed on to heirs, did not comprise an important fraction of planters' total assets, however much they may have caught the eye of visiting foreigners or Brazilian creditors.[15] Even luxury services, such as European vacations and schooling, represented only a moderate expense compared to the cost of new machinery. Between 1870 and 1881, for example, planters in Escada, one of the richest sugar counties, may have made an average of two trips per year. These trips would have cost between two and three contos each, or the value of a fair harvest of a traditional engenho. Such junkets obviously cost far more than the average senhor de engenho could afford: in fact only the sugar oligarchs, members of each country's richest families, so indulged. These luxuries did not distract capital from all productive investment, for some of those oligarchs were the only planters who succeeded in modernizing their engenhos in the nineteenth century without subsidies.[16]

14. Freyre, *The Masters and the Slaves*, p. 55. *Idem.*, Freyre, *The Mansions and the Shanties*, p. 119.

15. Koster, *Travels in Brazil*, v. I, pp. 67, 275-276. Uma das victimas, "Banco de Crédito Real," *Diário de Pernambuco*, June 1, 1887.

16. Our estimate derives from an annual examination of three months' of passenger lists published in the *Diário de Pernambuco*, where we found that five Escada planters sailed directly for Europe during those twelve years. If one assumes that other planters sailed as frequently in the other months, or a total of twenty sailings, and that few planters chose the much longer route via Rio de Janeiro, one arrives at a total under thirty for the county. Round trip first-class passage between Recife and Lisboa in 1874 cost 600$000 for the planter and his wife, and 150$000 for a child. It is harder to estimate expenses in Europe; if living, purchases and travel cost twice as much as the transatlantic passages, then the planter spent a total of 2:250$000 on himself, wife, and one child. Schooling, of longer duration, might cost more, but only the child's maintenance was involved.

Land was the planter's most valuable asset (Table 16). Since he worked only a fraction of his land, he could presumably raise cash by selling, renting, or at least mortgaging extra or unused property. Depending upon size and location, an average engenho was worth about 25 contos at the beginning of the century, and up to 200 contos at century's end. Rents around the 1870's and 1880's varied between one and three contos per year for a simple plantation and 1$000 per sugar loaf for a plantation with a mill.[17]

As in the case of slaves and livestock, however, sales and rentals of land alienated productive factors. The planter wishing to maintain or increase production, one would imagine, would prefer to mortgage. But such was not the case:

The planters and the people with whom they deal generally do not fund their business with mortgage contracts, but with the emission of letters of credit which have, among other advantages, that of mobilizing capital. The judges of law tell us now . . . that the repugnance for loans by contracts of that kind [mortgages] is many times unconquerable . . . [due to] the expenses which they occasion, or the publicity of the debt and the immobilizing of the goods.[18]

Both correspondentes and planters distrusted mortgages. During most of the colonial period, Portuguese law had prohibited foreclosures to encourage the sugar industry: creditors could claim only the sugar mill's income, not the property itself. In the nineteenth century foreclosures were permitted, but the creditor was obliged to accept the property, or the price it fetched in

17. Koster, Travels in Brazil, v. II, pp. 139, 195. Koster noted the plantations nearest the Recife market brought higher prices, probably due to transport costs. Archive of the Usina União e Indústria (Recife and Escada) File on Engenho Recreio, letter dated February 27, 1878. Ibid., File on Engenho Bomfim, letter dated June 5, 1882. Milet, A Lavoura da Canna de Assucar, pp. 105, 111. Idem., "Preços do Assucar e Futuro da Nossa Industria Assucareira," Diário de Pernambuco, December 16, 1886. "Engenho," ibid., February 1, 1888. While some affirmed that the end of the international slave trade in 1850 had caused engenho rents to drop, in reaction to increased slave prices, our materials reveal no such fall. "O dr. José Antonio de Figueiredo ao Público, XVIII," A Província, January 30, 1875.

18. Additamento às Informações sôbre o Estado de Lavoura (Rio de Janeiro, 1874), p. 43.

public auction, as full payment for the debt. A creditor feared loaning against a property whose auction price was uncertain, or, if no buyer appeared, "becoming a rural landowner against his wish." The law prohibited the sale of mortgages, so the forecloser had only these two alternatives.[19]

The Rio de Janeiro authorities tried without much success to reform mortgage law. In 1846, the imperial government had established a registry for conventional mortgages, but other forms of property transfers, such as sales and rentals, remained unregistered, with the result that when creditors came calling, the debtor was often hard to find. During the years 1860-64, when prime interest rates were declining and many mortgages could have been expected, the total average annual mortgage debt in Pernambuco amounted to 1,278 contos, the value of only 20 to 25 engenhos. Soon thereafter, the imperial legislature passed Law 1,237 on September 24, 1864, which set penalties for failures to register mortgages and required that collateral have a specified value.[20] But these reforms did not achieve the desired results. In order to exact additional security and exorbitant interest rates, creditors refused to comply with registration requirements. Moreover, the mortgage-holders had no obligation to erase foreclosed mortgages from the register. On the other hand, mortgagees who purchased an encumbered property were not required to erase the old debt but only to register their own loan. Because of such imperfect records, the same property was often mort-

19. Pereira da Costa, "Origens Históricas," p. 286. Koster, *Travels in Brazil*, v. II, pp. 131-132, SAAP, *Trabalhos do Congress Agrícola*, pp. 117, 168, 276. See also H. Christian Borstel to George Rives, Pernambuco, July 16, 1889, U.S. National Archives, Despatches from U.S. Consuls, v. 13 for a description of problems attending mortgages.

20. Brazil Congresso, Camara dos Deputados, *Reforma Hypothecaria* (Rio de Janeiro, 1856), p. 4. *Additamento às Informações sôbre o Estado de Lavoura*, pp. 43-44; Table 6, facing p. 46. These data contradict the customs inspector who estimated that two-thirds of the province's engenhos were mortgaged. He may have been referring only to those mills in a certain region. Report by Hunt, August 18, 1864, *Parliamentary Papers*, 1865, HCC, v. LIII, AP, v. XXIV, p. 351. Paes de Andrade. *Questões Econômicas*, p. 53. *Lei e Regulamento da Reforma Hypotecaria estabelecendo as bases das Sociedades de Credito Real* (Rio de Janeiro, 1865) pp. 7-12.

gaged several times, by father and then by son, by husband and then widow, and no creditor could be sure that his debt would receive first attention.[21]

Brazil legalized the Torrens Register by Decree 370 on May 2, 1890; this Register, originally instituted in Australia in 1859, recorded officially all property and owners and issued transferable property deeds which could be used to secure a mortgage. But in Brazil the Torrens Register did not gain acceptance, and the regulations specifying its implementation were not approved. Critics objected that the Register required the landowner to present public legal documents proving title. Since landowners often had acquired their properties illegally or without documents, under the Torrens Register they would have lost possession.[22] As a result of this resistance, by 1909 registered mortgages in Pernambuco totalled only 4,152 contos, the equivalent value of less than fifty engenhos.[23] The landowner's fear of exposure of prior debts and imperfect title defeated these attempts to broaden the capital market.

Associations

Given the difficulties that individuals had in mobilizing capital through mortgages, many Pernambucan entrepreneurs reached the obvious conclusion that such mobilizing might be accomplished better by an association of individuals. Unfortun-

21. *Additamento às Informações sôbre o Estado de Lavoura*, pp. 43-44. "De que precisa a industria," *O Industrial. Revista de Industrias e Artes* (Recife) Ano I, no. 2, February 15, 1883, pp. 15-16. For unsuccessful appeals for further reforms, see SAAP, *Trabalhos do Congresso Agrícola*, pp. 117, 176, 387, 441.

22. SAAP, *Trabalhos do Congresso Agrícola*, p. 122. "A Lei Torrens," *Jornal do Recife*, June 4, 1890. Luiz Souza Gomes, *Dicionário Econômico e Financeiro*, 8th ed. (Rio de Janeiro, 1966), p. 222. *Trabalhos da Conferência Assucareira*, Part 1, p. 93, Part 2, pp. 43, 45. For the text of the law, which also offered mortgages, see "Decreto 370," *Decretos do Governo Provincial da República dos Estados Unidos do Brazil, quinto Fascículo, de 1 a 31 de Maio de 1890*, pp. 798-847. The landowners also feared allowing women equal rights to property jointly held with husbands.

23. Direction Générale de Statistique, *Annuaire Statistique du Brésil*, 1ère Année (1908-1912), 2 vols. (Brésil, 1917), v. II, *Economie et Finances*, pp. 174-179.

ately for these capitalists, the laws of the last thirty years of the Empire did not encourage such activities. The later 1850's had witnessed a proliferation of banks of emission authorized by the Banco do Brasil. Credit and speculation expanded, until in 1857 the first of several crises occurred when foreign creditors in Europe and the United States, reacting to their own financial crises, demanded payments which drained specie from Brazil. New foreign loans and more paper emissions, both legal and unauthorized, compounded the problem, and business failures multiplied as major firms in capital cities declared bankruptcy.[24]

The inflation and speculation only ended when the imperial legislature approved Law 1,083 of August 22, 1860, which sharply restricted emissions by both private banks and the Banco do Brasil. Most important for the sugar industry, the law required that any limited liability company had to receive authorization from the Council of State before it could operate, and also had to submit annual reports for approval. By requiring business to seek such political approval, the law inhibited the corporate organization of capital. Pernambucan entrepreneurs unhesitatingly condemned the "restrictive and centralizing" 1860 law as "the beginning of the annihilation of the spirit of enterprise and association," and "the most perfect instrument to kill the spirit of association and individual initiative itself."[25]

Only 22 years later did new legislation begin to relax the 1860 law. New guarantees for investors and definitions of the responsibilities of company founders and officers were stipulated in Law 3,150 of November 4, 1882, which exempted all limited liability companies from government authorization, with the specific exceptions of religious organizations, charitable institutions, savings and emissions banks, mortgage and real estate companies, insurance companies, and companies dealing in food-

24. J. Pandiá Cológeras, A Política Monetária do Brasil, translated by Thomaz Newlands Neto (São Paulo, 1960), chapters viii-ix, passim.
25. "Lei 1,083," Colleção das Leis do Império do Brasil de 1860, tomo XXI, Parte I, pp. 28-36. SAAP, Trabalhos do Congresso Agrícola, p. 195. Ceresiades, "A agricultura em Pernambuco, IX, Bancos Ruraes," Diário de Pernambuco, July 5, 1878. Henrique Augusto Milet, O Meio Circulante e a Questão Bancária, 2nd ed. (Recife, 1875), p. 123, cited by Werneck Sodré, História da Burguesia Brasileira, p. 161.

stuffs. With the advent of the Republic, the lawmakers revised corporate legislation: Decree 164 of January 17, 1890, repeated most of the 1882 law, with the important nationalistic provisions that foreign firms had two years to raise two-thirds of their capital in Brazil and all stockholders resident in Brazil would be subject to Brazilian law. All companies had to publish annual lists of stock transactions.[26]

The 1882 and 1890 laws in effect permitted once more the freer functioning of most corporate enterprise. In particular, central mill and usina company directors no longer needed friends in Rio de Janeiro to guarantee their survival, and investors in those companies felt a modicum of confidence in those to whom they entrusted their capital. The proliferation of limited liability companies, including usina companies such as the Companhia Geral de Melhoramentos de Pernambuco, the Companhia Agrícola Mercantil de Pernambuco, the Companhia Florestal Agrícola, the Companhia Assucareira de Pernambuco, and the Companhia Progresso Colonial, in the early years of the Republic, testified to the efficacy of these reforms.

Banks

Many speakers at the 1878 Recife Agricultural Congress argued that the foundation of special mortgage banks would best augment the capital supply and circumvent the resistance against conventional mortgages. For most of the nineteenth century, the planters had received no help from the banks because few of these institutions lasted very long, and those that did preferred a commercial clientele. Thus commercial banks would loan at 8 or 9 percent for a month or two to Recife merchants and correspondentes, who would then reloan at higher rates to the planters. The money eventually reached the sugar industry, but a

26. "Lei 3,150," *Collecão das Leis do Imperio do Brasil de 1882*, Parte I, Tomo XXIX, v. I, pp. 139-149. "Decreto 164," *Decretos do Governo Provisório da República dos Estados Unidos do Brazil. Primeiro Fascículo de 1 a 31 de janeiro de 1890*, pp. 83-94. The significance of the 1882 and 1890 reforms is also traced briefly by Stanley J. Stein, *The Brazilian Cotton Manufacture, Textile Enterprise in an Underdeveloped Area, 1850-1950* (Cambridge, Mass., 1957), p. 8; Vilela Luz, *A Luta pela Industrialização*, pp. 39-40; and Graham, *Britain and the Onset of Modernization*, pp. 224-230.

shorter route would have been less expensive, and could have permitted capital loans.[27]

Between 1808 and 1880, at least eleven banks were founded in Recife, not counting other commercial houses engaged only in trading foreign exchange and discounting commercial paper. But hardly any offered rural mortgages.[28] Mortgage banks only became a real possibility in the final years of the empire, as a result of government subsidies. Only one such bank, however, reached the operational stage. The imperial government first began offering subsidies through Law 1,237 of September 24, 1864, which simply authorized mortgage societies to emit mortgage paper. The subsequent Law 2,687 of November 6, 1875, guaranteed a 5 percent return and amortization of thirty-year 7 percent mortgage loans by a Banco de Crédito Real which raised its capital in Europe.[29] But the Treasury minister complained in 1877 that the English capitalists refused to invest in the proposed bank because they feared that "the competition between the mortgage paper and our foreign debt bonds" would produce falls in the values of both.[30] To attract investors, the government subsequently increased the guaranteed return to 6 percent, and then to 7 percent, and allowed proposed banks to raise their capital in Brazil. Between 1879 and 1882 various entrepreneurs proposed founding banks in Pernambuco with these government guarantees, but no bank appeared. The ACA failed to establish

27. SAAP, *Trabalhos do Congresso Agrícola*, pp. 144, 176, 215, 219, 280, 441. Paes de Andrade, *Questões Econômicas*, pp. 52-53.

28. F. A. Pereira da Costa, "Notícia sôbre as instituições de crédito bancário em Pernambuco, offerecida à Benemerita Associação Commercial Beneficente," *Relatório da Direcção da Associação Commercial Beneficente de Pernambuco apresentado à Assembléa Geral da mesma em 10 de agôsto de 1898*, pp. 95-112. Pereira da Costa wrote the best review of nineteenth century banking in Pernambuco. Flávio Guerra, "A Arte de Amealhar Dinheiro," *Diário de Pernambuco*, May 7, 1967, Caderno V, pp. 10-11, borrows liberally from Pereira da Costa.

29. *Lei e Regulamento da Reforma Hypothecaria*, p. 11. Sociedade Nacional de Agricultura, *Legislação Agrícola do Brasil, Primeiro Período Império (1808-1889)* (Rio de Janeiro, 1910), v. II, Part 2, pp. 12-13.

30. *Proposta e Relatório apresentados à Assembléa Geral Legislativa pelo Ministro e Secretário do Estado dos Negócios Estrangeiros e Interino da Fazenda Barão de Cotegipe* (Rio de Janeiro, 1877), p. 24.

its Banco Commercial Agrícola e Hypothecario de Pernambuco in 1883 because the Emperor disapproved of its statutes. Nor could the SAAP launch its Banco Auxiliadora de Agricultura.[31]

The one and only successful mortgage bank in nineteenth-century Pernambuco, the Banco de Crédito Real de Pernambuco opened its door in Recife's financial district in 1886, during a crisis of low sugar prices and foreign exchange rates. Capitalized at a modest five hundred contos subscribed mostly by Recife merchants and sugar exporters, the Banco de Crédito Real began making loans in negotiable paper at 8 percent for up to thirty years against half the value of rural properties and three-quarters the value of urban properties. Three wealthy merchants, Manoel João d'Amorim, José da Silva Loyo Júnior, and Luís Duprat composed the Banco directorate.[32]

Despite its modest size, the Banco de Crédito Real immediately attracted enough borrowers to enrage the correspondentes, who formerly had monopolized rural credit. One correspondente wrote that "the planters were appearing to deceive the good faith of their correspondentes and ex-correspondentes with promises that they were going to mortgage their mills to pay them, so as not to cause them the least difficulty. But after the mortgage is made [to the Banco de Crédito Real de Pernambuco], and the mortgage paper sold for cash, they take to their heels, as it is said, leaving the correspondentes swindled, without anything to claim in payment, since the mill is the only valuable good they own." The correspondente asked the Banco directors to refuse mortgage applications by planters "unmindful of their creditors . . . who forget

31. *Annaes da Assembléa Legislativa Provincial de Pernambuco no Anno de 1882*, May 29, 1882. Various authors, "Banco de Credito Real," *Diário de Pernambuco*, May 20, August 25, 1882. "ACA," *ibid.*, May 18, 1883. "Banco Commercial Agrícola e Hypothecario de Pernambuco," *ibid.*, June 3-July 21, 1883, *passim*. Sociedade Auxiliadora da Agricultura de Pernambuco, *Estatutos para o Banco Auxiliador d'Agricultura. Instituição de Crédito Real e Agrícola* (Recife, 1883). "Banco Auxiliador d'Agricultura de Pernambuco," *Diário de Pernambuco*," August 19, 1883; December 2, 14, 20, 1883.

32. "Banco de Crédito Real em Pernambuco," *ibid.*, January 19, 1886. *Relatório do Banco de Crédito Real de Pernambuco apresentado à Assembléa Geral dos Accionistas em 23 de março de 1889* (Recife, 1889) pp. 33-37, gave the bank's stockholders and shares as of 1888.

all the favors done, sacrificing the purest sentiments of the heart, trampling underfoot the sacred duties of honor."[33]

A sugar planter replied through the *Diário de Pernambuco* that the correspondentes were discounting Banco de Crédito Real mortgage paper up to 10 percent, "and even this as a great favor."[34] The correspondente retorted that if the planter did not trust his correspondente he could sell his sugar elsewhere:

Yesterday, when there were no banks, to whom did they run mortgaging their mills? To the correspondentes, honorable merchants, lovers of agriculture. Nobody made an ugly face at their money, even at high interest rates. They accepted the loans contracted as no small favor. Nobody dared to say that the money was lent purposely to become creditors, or other such outbursts. Today, when the majority of the correspondentes have drained their strongboxes, exhausted their resources, the possibility exists of running to the Banco to mortgage sugar mills in order not to pay creditors. It is the fashion to make false mortgages, to swindle left and right. Today the mud of the streets is thrown at the class of correspondentes.[35]

With only five hundred contos capital, however, the Banco de Crédito Real could not make much of a contribution to modernizing sugar mills. The provincial assembly attempted to strengthen the Banco by granting tax exemptions on its operations and mortgage paper, and by making that paper acceptable collateral for provincial loans, but the provincial president, perhaps pressured by correspondentes, rejected these laws as unwarranted subsidies.[36]

In the wake of the abolition of slavery and in response to popular demand, the imperial government gave the Banco de Crédito Real the aid necessary to make a significant contribution to the sugar sector. To compensate for the loss of slaves, to increase cash

33. Uma victima, "Banco de Crédito Real," *Diário de Pernambuco*, May 21, June 1, 1887.

34. Uma victima das victimas, "Banco de Crédito Real," *ibid.*, May 28, 1887.

35. Uma victima, "Banco de Crédito Real," *ibid.*, June 21, 1887.

36. *Relatório da Direcção da Associação Commercial Beneficente de Pernambuco, apresentado à Assembléa Geral da mesma em 9 de agôsto de 1887*, p. 35.

available to meet payrolls, and to facilitate credit which formerly had depended in part on slave collateral, the imperial government loaned six thousand contos interest-free to the Banco do Brasil in Rio de Janeiro, São Paulo, Minas Gerais, and Espírito Santo. The Banco do Brasil had to match this amount, and then was authorized to loan the total twelve thousand contos at 6 percent. The ACBP and ACA immediately demanded that João Alfredo, the Pernambuco-born president of the Council of Ministers, extend a similar subsidy to the northern provinces. A few weeks later, the imperial government granted three thousand contos to the Banco de Crédito Real, which would loan the money at 6 percent to borrowers in Pernambuco, Paraíba, Rio Grande do Norte, and Alagoas. In deference to the poverty of northeastern capital markets, no matching funds were required.[37]

Invigorated by this subsidy, the Banco de Crédito Real quickly expanded its operations into financing usinas. The Banco offered mortgage loans worth one-half the value of the property and one-third the value of the sugar mill, as evaluated by the Banco. In less than one year, the Banco had financed five new usinas: Bamburral, Santa Philonila and Cabeça de Negro in Escada, Carassú in Barreiros, and Bandeira in Ipojuca. The Banco also helped improve the Usinas Bosque and Aripibú in Escada. These seven usinas could each produce an average of nine hundred tons of sugar per harvest, over six times the capacity of the traditional engenhos.[38]

The Banco de Crédito Real weathered the financial storms of the 1890's and regularly paid its stockholders 10 percent annual dividends. After twenty-five years of operation, the Banco reported that three-fourths of a total of 14,209 contos in long-term loans had gone to rural borrowers. Moreover, amortizations had reached 65 percent of the total. The Banco remained the only successful long-term agrarian credit institution in Pernambuco in

37. *Relatório da Direcção da Associação Commercial Beneficente de Pernambuco apresentado à Assembléa Geral da mesma em 12 de agôsto de 1888*, p. 25. Calógeras, *A Política Monetária do Brasil*, pp. 180-182. "SAA," *Diário de Pernambuco*, July 22, 1888. "ACA," *ibid.*, July 28, 1888. "Auxilio a lavoura do norte," *ibid.*, September 12, 1888.
38. *Relatório do Banco de Crédito Real . . . 23 de março de 1889*, p. 11.

the nineteenth century. The most stable commercial banks, the English branches, avoided direct dealings with planters, "egotistically limiting themselves to the game of exchange rates."[39]

Government

The unique success of the Banco de Crédito Real derived from its imperial subsidy. Pernambuco planters had long recognized the government as their most powerful ally in the struggle to increase the supply of credit. The government had basically three ways of helping: it could subsidize and guarantee regional banks to improve their capacity and inclination to lend to agriculture, as in the case of the Banco de Crédito Real; it could simply increase the supply of money; or it could grant direct subsidies.

Since the imperial government had occasionally resorted to monetary policies in times of national or regional crisis, many senhores de engenho demanded similar policies when their sugar economy experienced hard times (Table 17).[40] One year after abolition the ministry of the Visconde de Ouro Preto tried to recover political support for the monarchy by making a total of 172,000 contos available to regional banks for agricultural credit.[41] The Banco do Brasil authorized the commercial firm of Perei-

39. "Banco de Crédito Real," *Diário de Pernambuco*, August 8, 1894; March 31, 1897; July 20, 1899. *Relatório . . . ACBP . . . 9 de agôsto de 1897*, pp. 37-38. *Relatório . . . ACBP . . . 8 de agôsto de 1899*, p. 8. Banco de Crédito Real de Pernambuco, *Relatório apresentado à Assembléa Geral dos Accionistas em 29 de maio de 1912* (Recife, 1912), table facing p. 12. *Relatório da Direcção da Associação Commercial Beneficente de Pernambuco apresentado à Assembléa Geral da mesma em 6 de setembro de 1890*, p. 16. The British Consul described his countrymen's banks less angrily: "British banks are not permitted by their London Board of Directors to invest in industrial native enterprises." Arthur L.G. Williams, "Report for the Year 1894 on the Trade of the Consular District of Pernambuco," *Parliamentary Papers*, 1895, HCC, v. XCVI, *AP*, v. 36, p. 3. David Joslin, *A Century of Banking in Latin America:toCommemorate the Centenary in 1962 of the Bank of London and South America, Limited* (London, 1963), cited in Graham, *Britain and the Onset of Modernization*, pp. 97-98.

40. SAAP, *Trabalhos do Congresso Agrícola*, p. 356, 439. João Fernandes Lopes, "Publicações a Pedido. Questões economicas," *Diário de Pernambuco*, January 11-12, 1887.

41. Calógeras, *A Política Monetária do Brasil*, pp. 193-194. "Restrospecto Commercial do anno de 1889," *Diário de Pernambuco*, January 12, 1890.

TABLE 17

BRAZILIAN PAPER MONEY EMISSIONS

Years	Average Annual Emissions (contos)	Probable Cause for Increase
1847-50	49,152	
		New emission banks backed by capital formerly employed in slave trade
1856-60	92,912	
		Paraguayan War
1865-70	146,130	
		Severe droughts in Northeast
1877-80	213,842	
		Encilhamento and new emission banks
1891-98	671,235	

SOURCE: Ónody, A Inflação Brasileira, pp. 27-28.

ra Carneiro e Cia. in Recife to handle the loans to northern provinces. Pereira Carneiro offered 6 percent loans against property, equipment, harvests, stocks, and securities. But the Banco do Brasil did not permit loans larger that ten contos per borrower for periods longer than two years; moreover, the delinquent borrower incurred a penalty interest of 9 percent, plus expenses. These small, short-term loans led one senhor de engenho to call them "ridiculous . . . impossible for the greater number of our planters," and another regretted that "the aid is only to commerce, upon which agriculture continues to depend."[42] The absence of any records documenting how this money was loaned in Pernambuco suggests that the terms imposed and the subsequent fall of the Empire in November 1889 prevented distribution.

42. Henrique Augusto Milet, "O preço de assucar e os Bancos do Conselheiro Ruy Barbosa," ibid., February 23, 1890. Paulo d'Amorim Salgado, "A Crise Agrícola e arbitros para sua solução," ibid., May 2, 1897.

Under the early Republic, several emission banks increased the volume of currency circulating in Perambuco, and contributed to the giddy inflation and speculation known as the *encilhamento*. After multiple emissions were prohibited, some individuals and municipal governments began issuing their own numbered and signed currency, "which circulates according to the credit of each issuer." Opinions were sharply divided on the utility of emissions. When the federal government reduced money in circulation, protests quickly appeared in the Recife press.[43] On the other hand, others argued that emissions would cheapen the mil-réis—that is, raise the price of foreign exchange—and thereby make modernization with imported machinery more costly. Several merchants objected that emissions earmarked for agriculture would constitute a forced loan from other groups, since the general price level would rise while only planters would enjoy increased purchasing power. Others lamented that such a policy would start general price rises at the end of each harvest, when crops reached markets, with the result that by the beginning of the next harvest cash would be as scarce as before. The fall of the mil-réis to nearly one-fourth of par, during the early Republic, confirmed the worst of these fears, although it also inflated export revenues and stimulated the sugar industry.[44]

43. O velho aldeião, "A lavoura e os lavradores," *Jornal do Recife*, July 21, 1894. Henrique Augusto Milet, "Auxiliadora da Agricultura," *Diário de Pernambuco*, April 27, 1894. "Plano de resgate," *ibid.*, January 27, 1897. O matuto, "A crise agrícola," *ibid.*, May 5, 1897. Um agricultor, "A crise agrícola," *ibid.*, May 8, 1897.

44. SAAP, *Trabalhos do Congresso Agrícola*, pp. 134, 118. Henrique Augusto Milet argued that the exchange rate depended upon the demand for foreign currency, not the amount of Brazilian currency in circulation. If that demand remained constant, then business activity would increase without a price rise. *ibid.*, pp. 151-153. In essence, Milet was anticipating the Keynesian criticism of the classical quantity theory of money as formulated by Fisher. Alfred W. Stonier and Douglas C. Hague, *A Textbook of Economic Theory* (New York, 1961), pp. 473-478. But he forgot that modernization entailed machinery imports and hence an increase in the demand for foreign exchange. Affonso de Albuquerque Mello, "Os bancos de emissão do Sr. Fernando Lopes," *Diário de Pernambuco*, January 12, 1887. *Idem.*, "Bancos de emissão, IV" *ibid.*, August 10, 1887. "A baixa do cambio," *ibid.*, March 16, 1897. Fernando de Castro, "A

If emissions threatened monetary stability, there still remained the alternative, often resorted to by the imperial government, of borrowing abroad. By the Constitution of 1891, the state governments retained the right to borrow abroad. At least one prominent planter suggested such loans in lieu of additional state bond emissions.[45] But several factors militated against such a policy in Pernambuco. First, the state governments took very seriously their indebtedness from paying interest on bonds emitted to subsidize usinas: constant references to that debt in the state governors' annual messages imply a fear of incurring still greater obligations from more exigent creditors. Secondly, the general downward trend of the foreign exchange rate naturally aroused fears that amortizing foreign loans might prove far more costly than anticipated. Moreover, even if the state government resorted to a special tax on exports to finance amortization, as did the coffee states in the Taubaté Convention of 1906, the poor prospects of sugar exports in the 1890's, as well as the political price of increasing taxes on the elite, discouraged relying on such a measure.[46]

Neither bank subsidies nor emissions nor foreign loans proved very successful in promoting the modernization of the sugar industry. To protect debtors and investors, the traditional financial infrastructure placed strong restraints on mobilizing capital; even when these restraints were partly relaxed, private initiative responded slowly. Only when the government directly subsidized central mills and usinas did the pace of modernization accelerate.

---

crise da lavoura," *ibid.*, May 12, 1897. C. C., "A crise agrícola," *ibid.*, June 5, 1897. Allan Paterson, "Movimento agrícola," *ibid.*, May 8, 1901.

45. José Ruffino, "Concessões de Usinas," *Jornal de Recife*, August 6, 1895.

46. Henrique Augusto Milet, "O preço de assucar," *Diário de Pernambuco*, September 27, 1890. "Emprestimo externo," *ibid.*, January 3, 1899. Fernando de Castro, "A crise da lavoura . . ." *ibid.*, May 12, 1897. The coffee producers, who dominated the world market, could transfer part of the tax to the consumers.

5

CENTRAL MILLS AND
USINAS: A SUBSIDIZED
MODERNIZATION

𝒢overnment subsidies facilitated the principal efforts to mo-
bilize capital and modernize the Pernambuco sugar economy in
the later nineteenth century. These subsidies all had one goal: to
promote the establishment of large, modern sugar mills. The
subsidies took three principal forms: subsidies to local credit
institutions, as described above; guaranteed returns on investments; and capital loans in government bonds.

Central Mills

The central mill meant fundamentally a division of labor: the
plantation owners growing the sugar cane, instead of processing it
themselves, would send it to neighboring central mills. This
arrangement, it was hoped, would make more efficient use of productive factors: each plantation could devote its resources solely to
agriculture, thus facilitating improvements such as the use of fer-
tilizers, irrigation, and mechanization. The planter would not
need to maintain and improve his mill; in fact, he could abandon
it. The central mill owners, on the other hand, could devote all
their resources to the industrial sector, improving the mill
through capital investments. They would not need to invest in
cane growing. The results would be cheaper cane and cheaper,
more competitive sugar.

The central mills would encourage a more economical use of
land and labor. The cane growers would have sufficient capital to
work more of their land. The mill owners would expand mill capacity with their increased capital and would demand more cane.

The result would be more area under cultivation. Central mills also could encourage a cheaper labor supply. Each plantation owner's total labor demand would decline by the number of workers discharged from the mill and not reemployed in growing cane. Similarly, each central mill owner would employ only enough workers to operate his mill. In addition, both planter and miller would use part of the money saved to make technological improvements, which in many cases were capital-intensive and labor-saving. The workers rendered idle by these changes would then exert downward pressure on the wage level.[1] In fact, the wage level did decline after the middle 1870's, when the first modernized mills appeared, although it would be difficult to establish a direct causal relationship (see Chapter Eight).

For these reasons, most Pernambucans nourished great hopes for central mills in the 1870's. A provincial president characterized central mills as "the saving remedy," and "the salvation of the sugar industry." Planters styled them "the life raft to which we must cling," and called for "the profitable introduction of central factories, with their powerful machines and perfected apparatus" which would introduce "a true revolution in the work pattern." An ACA president hoped for central mills "in every judicial district, in any parish, in any location."[2]

But for all the advantages and all the enthusiasm, Pernambucans had to wait over twenty-five years before the first successful central mill began operation. As early as 1857, Ignácio de Bar-

1. Abolitionists expected that the central mills' demand for specialized labor, and the concommitant reduction in demand for unskilled workers, would help end slavery. Graham, *Britain and the Onset of Modernization*, p. 150.

2. *Falla com que o Exm. Sr. Dr. Adolpho de Barros Caval^te de Lacerda Presidente da Província Abrio a Sessão da Assembléa Legislativa em 19 de dezembro de 1878*, p. 60. *Falla com que o Exm. Sr. Dr. Adolpho de Barros Cavalcante de Lacerda Presidente da Província abrio a sessão da Assembléa Legislativa em 1º de março de 1879*, p. 21. Milet, *Os Quebra Kilos*, p. 56. SAAP, *Livro de Atas* no. 2, August 24, 1875. See also *idem.*, *Trabalhos do Congresso Agrícola*, pp. 306, 378, 389. The principal of the central mill was not new. Since colonial times engenhos had acquired cane grown by sharecroppers on the plantation, in exchange for residence rights and a share of the sugar made from that cane. On occasion the engenhos even bought cane from other plantations run by individuals without the means to operate a mill. Canabrava, "A Grande Propriedade Rural," p. 207.

ros Barreto appealed to the Pernambuco provincial assembly to aid a French central mill company represented in Brazil by Charles Louis Richard de Lahautière. The provincial assembly sanctioned expropriation in the public interest, so that the projected mill could carry its cane over other people's property. But Lahautière, unable to conclude contracts with local sugar cane suppliers, requested several postponements of the contract, and finally returned discouraged to France.[3]

Despite this failure, Pernambucans remained interested in central mills. During the 1860's, several authorities on sugar production in Brazil published books recommending central mills. The successful experiences in the French West Indian islands of Martinique and Guadelupe were impressive: one entrepreneur reported that the French colonial mills were averaging 25 percent annual profits, and on Martinique profits were as high as 31 percent. Admiring references to British-built central mills in Egypt appeared in the 1870's. Planters marveled that only seventeen Egyptian central mills produced nearly as much sugar as all 1,500 Pernambucan engenhos in 1873.[4]

3. Ignacio de Barros Barreto, "A Dignissima Assembléa Legislativa de Pernambuco e o Acoroçoamento da melhoraria do Fabrico de Assucar," *Diário de Pernambuco*, May 27, 1876. "Engenhos Centraes," *O Brasil Agrícola*, Ano III, February 28, 1882, p. 91. Ignacio de Barros Barreto to Epaminondas de Barros Correia, *ibid.*, March 15, 1882, p. 101. Pereira da Costa, "Origens Históricas," p. 304.

4. Ferreira Soares, *Notas Estatísticas*, pp. 103-105. Burlamaque, *Monographia da Canna d'Assucar*, p. 67. Paes de Andrade, *Questões Econômicas*, pp. 88, 91. Major Anfrísio Fialho, *Impending Catastrophe. The C-S-F-of-, Ltd.* (London, 1884), p. 2. "Engenhos Centraes," *Diário de Pernambuco*, November 11, 1875. Partido Conservador, "Agricultura," *ibid.*, February 11, 1875. *Relatório . . . ACBP . . . 6 de agôsto de 1875*, article XXVIII. *Relatório apresentado à Assembléa Geral Legislativa na Primeira Sessão da Décima Sétima Legislatura pelo Ministro e Secretário dos Negócios da Agricultura. Commércio e Obras Públicas João Lins Vieira Cansansão de Sinimbú* (Rio de Janeiro, 1879), p. 31. *Relatório apresentado à Assembléa Geral na Terceira Sessão da Décima Sétima Legislatura pelo Ministro e Secretário de Estado dos Negócios da Agricultura, Commercio e Obras Públicas Manoel Buarque de Macedo* (Rio de Janeiro, 1880), p. 132. The Ten Years' War (1868-78) delayed the appearance of central mills in Cuba until the late 1880's. Between 1885 and 1894, however, the number of Cuban mills dropped from 1,400 to 400, while production rose from 630,000 to 1 million tons. Many of the new *centrales* were founded by U.S. capitalists. Philip S. Foner, *A History of Cuba and Its Relations with the United States*, 2 vols. (New York, 1962-63). v. II, *1845-1895*, pp. 294-295.

During the 1870's, both the provincial and the imperial governments began offering subsidies for central mills. Despite various revisions in the form of these subsidies, however, none met with success. The provincial assembly passed Law 1,141 on June 8, 1874, which allowed provincial president Henrique Pereira de Lucena to promote six central mills by guaranteeing a return of 7 percent on five hundred contos per mill for twenty years. "Equal to those existing on the French islands of Martinique and Guadelupe," the projected mills would produce at least five hundred English tons (454 metric tons) yearly, three times the production of the traditional engenho. The Swiss Keller Company won the first contract awarded under Law 1,141 in January 1875. The company would build a mill in Água Preta in two years; it signed cane supply contracts in April 1875. The French sugar machinery firm Fives Lille received a similar concession in Cabo a few months later.[5]

The imperial government soon offered its own subsidies for central mills. Law 2,687 of November 6, 1875, guaranteed 7 percent for twenty years on 30,000 contos "to companies which propose to establish central mills [and] use the most perfected apparatus and modern processes." The company had to reserve 10 percent of its guaranteed capital for 8 percent loans to cane growers, and had to accept as security harvests, tools, "and any other object not included in mortgage contracts."[6]

Opposition to the original Keller contract arose when SAAP members charged that traditional mills refurbished with modern machinery could yield higher sugar cane ratios than the projected

5. "Lei 1,141," *Colleção de Leis Provinciaes de Pernambuco, Anno de 1874*, p. 58. Peres and Peres, *A Industria Assucareira em Pernambuco*, pp. 61-66. *Relatório com que o Desembargador Henrique Pereira de Lucena passou a administração desta província ao Comendador João Pedro de Carvalho de Moraes em 10 de Maio de 1875*, section "Engenhos Centraes." *Falla com que o Exm. Sr. Commendador João Pedro Carvalho de Moraes abrio a sessão da Assembléa Legislativa Provincial em o 1° de março de 1876*, pp. 100-101.

6. SNA, *Legislação Agrícola*, v. II, Part 2, pp. 12-14. José Honório Rodrigues, "A Revolução Industrial Açucareira, I," p. 84, emphasizes that since the central mill concessionaire would receive capital at 7 percent and loan to the planters at 8 percent, "as always the concessions to the Brazilian capitalist or the English capital exporter were higher than those to the planters and cane suppliers."

Swiss mill, which could profit only if it paid low cane prices. Due to this pressure and the imperial government's offer to guarantee profits in Law 2,687, the Keller contract was revised in February 1876. The principal changes included more safeguards for the government: the company's guarantee was doubled, but so was its required production. To prevent the company from inflating expenses to increase its guarantee earnings, only cane costs, raw materials, salaries, and repairs could be considered costs; new construction and replacements would be paid from a reserve fund. But Keller never began construction, allegedly because the company director died.[7]

Two Pernambuco senhores de engenho, the Barão de Campo Alegre and the Barão de Guararapes, received a similar concession in June 1876 for a central mill in Cabo, but they failed to use it. Fives Lille won an extension in 1877 and a revision in 1879, and went so far as to sign cane-supply contracts in Palmares. But "from then on nothing more was done"; the company failed to raise capital and the chief engineer died. In 1881, the provincial government annulled all outstanding concessions.[8]

These early attempts to subsidize central mills failed for two reasons: lack of capital and bad faith. The Agriculture minister

7. Ignacio de Barros Barreto, "Relatório de 1876 sobre o fabrico de assucar em Pernambuco," SAAP, Boletim, Fascículo no. 1 (1882), pp. 4, 57, 61. SNA, Legislação Agrícola, v. II, Part 2, pp. 20-25. Peres and Peres, A Industria Assucareira em Pernambuco, p. 66. Anfrisio Fialho, "Agricultura," Diário de Pernambuco, March 3, 1882.
8. SNA, Legislação Agrícola, v. II, Part 2, pp. 75-80. Falla . . . Adolpho de Barros Cavalcante de Lacerda . . . 1º de março de 1879, Anexo I, pp. 38-89. Falla com que o Excm⁰ Sr. Dr. Franklin Américo de Menezes Doria abriu a sessão da Assembléa Legislativa Provincial de Pernambuco em 1º de março de 1881, p. 94. The Fives Lille contract is available in "Contracto que fazem de uma parte," Pernambuco, Arquivo Público do Estado (APE), Coleção Engenhos Centrais, September 17, 1879. "Engenhos Centraes," O Brazil Agrícola, p. 91. Relatório com que o Excm⁰ Sr. Conselheiro Franklin Américo de Menezes Doria passou a administração d'esta provincia ao Excm⁰ Sr. Dr. José Antonio de Souza Lima em 7 de abril de 1881, p. 8. "Indice Geral e Chronologico das Leis e Decretos Relativos aos diversos serviços a cargo do Ministerio da Agricultura, Commercio e Obras Públicas até o Fim do Anno de 1887," Annexos ao Relatório apresentado à Assembléa Geral na terceira sessão da Vigesima Legislatura pelo Ministro e Secretario de Estado dos Negocios da Agricultura, Commercio e Obras Publicas Rodrigo Augusto da Silva (Rio de Janeiro, 1888), Appendix A, p. 98.

regretted that "the depression in money markets between 1876 and 1879 impeded the organization of enterprises based on guaranteed interest, due to the scarcity of cash and capital flight." It is not clear, moreover, why Keller or Fives Lille could not have raised capital in Europe, as did later English central mill concessionaires. The deaths of individuals should hardly have prevented corporate organizations from pursuing their purposes. The Agriculture minister did not mince words: "The lack of competence on the part of some concessionaires sterilized their concessions. They had sought these concessions purely for commercial speculation, hoping to transfer them to third parties who, capable of inspiring confidence, could raise the necessary capital through their commercial relations."[9]

These early failures, and more specifically the reality of the first cane contracts, modified the enthusiasm of some planters. Now they complained that becoming cane suppliers would place them "in conditions of inferiority," and would be "prejudicial to agriculture, abusing its state of decadence and inexperience."[10] Several insisted that central mills were only desirable when established by groups of neighboring planters, with their own cane and their own administration.[11]

In the 1880's, central mills became realities, although with considerable difficulties. The imperial government amplified Decree 2,687 with a new regulation, Decree 8,357 of December 24, 1881. This law tied the guaranteed capital to sugar production: 500 contos guaranteed for mills producing 1,000 tons, 750 contos guaranteed for twice that production, and 1,000 contos for

9. All quotes from the minister's *Relatório* for 1881, in Jeronimo de Viveiros, 'O Açucar através do periódico 'O Auxiliador da Industria Nacional'," *Brasil Açucareiro*, v. XXVII, no. 4 (April 1946), p. 408. Anfrisio Fialho, *Um Terço de Século (1852-1885) Recordações* (Rio de Janeiro, 1885), p. 150.

10. *Annaes da Assembléa Legislativa Provincial de Pernambuco, Anno de 1875*, session of May 14, 1875. See also "Collaboração, Breves considerações sobre a agricultura no Brasil. V," *Diário de Pernambuco*, March 23, 1876. SAAP, Livro de Atas no. 1 (Assembléa Geral), May 3, 1875.

11. Milet, *Os Quebra Kilos*, p. 57. *Idem., Auxilio à Lavoura e Crédito Real*, p. 5. *Engenhos Centrais no Brasil*, translated by Pereira Lima (Rio de Janeiro, 1877), p. 7. *Noticia acerca da Industria Assucareira no Brazil* (Rio de Janeiro, 1877), pp. 3-4. These latter two pamphlets reprint a speech by a General Morin to the Société Centrale de la Agriculture de France.

quadruple. Concessionaires would enjoy expropriation rights to facilitate cane transport, import tax exemptions, and preference in the use and purchase of public lands. The new law specified the distribution of the 30,000 contos, and Pernambuco received the largest share, over one-fourth the total (Table 18).[12]

An English company, the Central Sugar Factories of Brazil, Limited (CSFB), built the first central mills to operate in Pernambuco; but these mills performed miserably. Anfrísio Fialho, a lawyer, and Theodore Christiansen, an Englishman married to

12. "Decreto 8,357," *Colleção de Leis do Imperio do Brazil de 1881*, Parte II, tomo XLIV, v. II, pp. 1,387-395.

TABLE 18

ALLOTTMENT OF FUNDS UNDER IMPERIAL DECREES

SUBSIDIZING CENTRAL MILLS

Province	Capital Guaranteed			Share of Total (percent)		
	1881	1888	1889	1881	1888	1889
Pernambuco	8,000	7,500	6,400	26.7	25.0	21.3
Bahia	6,100	6,000	6,100	20.3	20.0	20.3
Rio de Janeiro	5,600	5,000	5,000	18.7	16.7	16.7
Sergipe	2,000	3,000	2,275	6.7	10.0	7.6
São Paulo	1,900	1,900	2,000	6.3	6.3	6.7
Rio Grande do Norte	1,500	1,400	1,000	5.0	4.7	3.7
Alagoas	1,200	1,200	2,275	4.0	4.0	7.6
Paraíba	700	900	1,150	2.3	3.0	3.8
Ceará	700	400	400	2.3	1.3	1.3
Maranhão	700	1,500	1,550	2.3	5.0	5.2
Pará	700	400	400	2.3	1.3	1.3
Espírito Santo	500	400	400	1.7	1.3	1.3
Município Neutro	400	300		1.3	1.0	
Minas Gerais		100	900		0.3	3.0
Totals	30,000	30,000	30,000 [a]	100	100	100

[a] Typographical errors presumably account for the missing 150 contos.

SOURCES: "Decreto 8,357," *Colleção de Leis do Império do Brazil de 1881*, Parte II, Tomo XLIV, v. II, p. 1,395. "Decreto 10,100," *Colleção de Leis do Império do Brazil de 1888*, Parte II, Tomo LI, v. II, p. 474. "Decreto 10,393," *Colleção de Leis do Império do Brazil de 1889*, Parte II, Tomo LII, v. II, p. 489.

the daughter of a wealthy planter, represented the CSFB in Brazil. In March 1881 they received their first concession, which guaranteed 7 percent on 1,500 contos to build three central mills. In October the first concession was increased to 2,100 contos, and they received a second guaranteeing 6 percent on another 2,100 contos to build three more mills. Thus the enterprising CSFB agents cornered over one-half of Pernambuco's allottment under Decree 8,357. Each mill had to produce 2,600 tons at a sugar-to-cane ratio of 5 percent.[13]

Fialho's cane contracts provoked a storm of controversy and the implacable hostility of the SAAP. Critics argued that the contracted cane price, 8$000 per ton, fell at least one-third below the value of sugar made from it, and well below cane and sugar beet prices in Spain and Germany. Others insisted that cane prices should depend not only on gross weight, but also on syrup density, the foreign exchange rate, or saccharine content. SAAP officers quickly labeled central mills "an economic mistake," and "premature, inopportune, and inexpedient." Antonio Gomes de Matos, a Recife merchant and abolitionist, not only criticized the cane price but also condemned the modern machinery: triple-effect vacuum pots were "a foolishness" more appropriate for beet sugar manufacture. He warned that the central mill guarantees increased the public debt, and feared that Brazil would be "governed by its creditors," as was Egypt, where British companies had also installed central mills.[14]

13. The two agents were in turn aided by Pernambuco senator and senhor de engenho Luis Felipe de Souza Leão, and the Brazilian ambassador in London. Fialho, Um Terço de Século, p. 167. Graham, Britain and the Onset of Modernization, pp. 150, 152. SNA, Legislação Agrícola, v. II, Part 2, pp. 202, 243-244.
14. Anfrisio Fialho, "Agricultura. Engenhos Centraes, Pernambuco," Jornal do Recife, January 27, 1881. Henrique Augusto Milet, "A Lavoura da canna de assucar relativamente à divisão do trabalho," SAAP, Boletim, September 1882, p. 9. This issue of the SAAP Boletim included many articles criticizing Fialho, the CSFB, and central mills in general. "Engenhos centraes," Diário de Pernambuco, December 16, 1882. For Fialho's replies and debate with another concessionaire, see "Engenhos Centraes," ibid., August 30, December 17, 1882. Ibid., March 3, 1882. "Jovino Bandeira e o Dr. Afrisio Fialho," Jornal do Recife, March 5, 7, 1882, and Diário de Pernambuco,

Having arranged concessions and cane contracts, Fialho transferred his rights to the CSFB in return for expenses, £50,025 in company stock, and the position of Adjunct Director and Company Representative in Brazil. But in 1883 the company fired Fialho and paid him £1,000 plus £3,000 for his stock—in all about 45 contos. The embittered promoter denounced his replacement for perpetrating "a premeditated swindle," and condemned Emperor Pedro II as "the worst enemy of material improvement" for having imposed onerous conditions on the concession.[15]

The ill will generated by the cane-price debates and Fialho's maneuvers increased as more evidence of bad faith came to light. Instead of new machinery as required by the concession, the CSFB installed rusted machinery bought used from British central mills in Egypt. During trials at the Cuyambuca mill in Água Preta in November 1883, the machinery broke, damaging turbines and injuring eight workers, one of whom later died. The breakdown delayed milling and left ripe cane in the fields. The CSFB offered to pay damages of 10 percent of the unmilled cane's value, but the planters, who had sold or abandoned their own machinery, remained furious. When the mill roof fell in a

---

March 7, 8, 1882. Quotes in SAAP, Livro de Atas no. 2, February 21, November 23, 1882; "Parecer sobre a conveniencia dos engenhos centraes," SAAP, *Boletim*, September 1882, p. 18. Antonio Gomes de Mattos, *Os Engenhos Centraes e o Sr. A. G. de Mattos, Colleçao dos artigos publicados no Jornal do Commércio de julho à agosto de 1882* (Rio de Janeiro, 1882), pp. 9-11. *Idem.*, *Esboço de um Manual para os Fazendeiros de Assucar no Brazil* (Rio de Janeiro, 1882), pp. xii-xiii. The critics also did not like the company's method of selecting cane to be cut, the delivery time, and the exclusion of certain parts from paid weight. But the cane price was the crux of the debate.

15. Fialho, *Um Terço de Século*, pp. 153, 172, 177-189. *Idem.*, *Impending Catastrophe*, pp. 2-4. Fialhos's political allegiances fluctuated considerably. In 1876 he had written for European readers a laudatory biography of the Emperor, in which he compared Pedro II to Trajan, Vespasian, Marcus Aurelius, Charlemagne, and other luminaries. Anfrisio Fialho, *Dom Pedro II. Empereur du Brésil, Notice Biographique* (Brussels, 1876), pp. 94-100. Nine years later he wrote a pamphlet calling for a republic in Brazil, which manuscript he then offered to sell to the Emperor for 60 contos. *Idem.*, *Um Terço de Século*, p. 187.

few months later and injured three workers, all hope of milling cane ended for that harvest season.[16]

The CSFB mill Bom Gôsto at Palmares also began inauspiciously. The mill did not grind cane when expected. One of its walls cracked and another was out of plumb. An old dam supplying water broke repeatedly, and prevented a single regular day of milling. The CSFB again offered the cane suppliers damages, but such compensation could not solve all problems. Payroll delays in 1886 so enraged a Brazilian worker that he killed a French assistant manager. The continual aggravations drove the English mill manager to drink, and he was removed in an advanced stage of delirium tremens.[17]

Even when the CSFB mills ground cane, the cane suppliers complained. The number of railway spurs was inadequate, the railroads made irregular pick-ups, let much cane fall on the ground, and often broke down. The mills made unwarranted deductions from cane weights, did not keep accurate scales, and did not provide contracted credit. Distillery wastes polluted the rivers and killed cattle.[18]

16. Francisco do Rego Barros to provincial president, October 31, 1883, and December 19, 1884, in APE, Coleção Engenhos Centraes. Various planters, "The Central Sugar Factories of Brasil. Protesto," *Diário de Pernambuco*, December 19, 1883. "Engenhos Centraes," and "Fabrica Central Cuyambuca," *ibid.*, February 8, 1884. Fialho, *Impending Catastrophe*, p. 30. "The Central Sugar Factories of Brazil," *O Industrial*, anno I, no. 11 (November 15, 1883), pp. 121-122. "Desabatamento do engenho central de Cuyambuca," *Diário de Pernambuco*, February 5, 1884.

17. The Bom Gôsto mill had not been included in the original concession, but on July 19, 1884, the imperial government had permitted the CSFB to substitute this site in exchange for a decrease in the guarantee to 6 percent. SNA, *Legislação Agrícola*, v. II, Part 2, pp. 304-305. "Fabrica central de Cuyambuca," *Diário de Pernambuco*, February 8, 1884. "Central Sugar Factories of Brazil Limited," *ibid.*, September 25, 1884. "Assassinato," *ibid.*, December 16, 1886. Francisco do Rego Barros to provincial president, February 26, 1886, APE, Coleção Engenhos Centraes.

18. Letters from Francisco do Rego Barros, November 19, 21, 1884; December 12, 16, 22, 1884; February 11, 1885; January 22, 1886; February 26, April 15, November 17, 1886, all in APE, Coleção Engenhos Centraes. Camara Municipal de Escada to provincial president, October 15, 26, 1884; March 5, 1886, in APE, Coleção Camaras Municipais, v. 66. "Relatório apresentado ao Ministério dos Negocios da Agricultura, Commercio e Obras Publicas pelo Engenheiro Fiscal Francisco do Rego Barros," *Annexos ao Relatório apresentado à*

The CSFB put four central mills into production for the 1884-85 and 1885-86 harvests. The mills disappointed all expectations. No mill extracted more syrup than traditional engenhos, only two exceeded significantly customary sugar-to-cane ratios, and none produced even one-half the sugar stipulated in the concessions. All ran sizeable deficits (Table 19).

The CSFB mills failed because the company did not want to succeed. Both English and Brazilian principals showed more interest in short-run profits than in long-run production. "For the same reason which the original concessionaire swindled the English, so these did the same to the government," editorialized one newspaper.[19] Another writer characterized the CSFB general manager as "one of the Arabians, who will do everything so that he and the other *innocents*, fellow countrymen and colleagues, eat the Oyster, leaving the Shell for the unsuspecting of Brazil."[20] A third charged the company with inflating expenses to increase guarantee payments, and an imperial official confirmed the charge.[21] More diplomatically, provincial officials and others

---

Assembléa Geral na Primeira Sessão da Vigesima Legislatura pelo Ministro do Estado dos Negócios d'Agricultura, Commercio e Obras Publicas Antonio da Silva Prado (Rio de Janeiro, 1886), Appendix A, pp. 3-5. O Matuto, "Engenhos centraes," and Algums agricultores logrados, "Escada, aos poderes publicos geraes e provincais," both in *Diário de Pernambuco*, July 10, 1886. "Engenho Central do Cabo," *ibid.*, July 13, 14, 1886. Various planters, "The Central Sugar Factories of Brazil Limited: Escada," *ibid.*, August 6, 1886. The Imperial Inspector-Engineer Rego Barros, who reported to the provincial president and the Agriculture minister, tried hard to be objective; not every complaint applied to every mill, and occasionally he rejected complaints as unwarranted. In general, however, he supported the criticisms.

19. "Engenhos centraes," *ibid.*, February 8, 1884.

20. O inquilino do hotel de Mondego, "Aos agricultores: Engenhos Centraes," *ibid.*, May 16, 1884. Emphasis in original.

21. Tiburcio de Magalhães, "Industria Saccharina, II," *ibid.*, December 18, and 19, 1883. "Relatório . . . Engenheiro Fiscal Francisco do Rego Barros," 1886, p. 5. Provincial president's representative to Minister of Agriculture, December 29, 1885, APE, Coleção Ministério da Agricultura, v. 12. The inspector refused to approve a marketing contract between the CSFB and a Recife exporter, because the company was already maintaining offices in Recife and London which could serve that purpose. He also feared that the company would sell cheaply to the exporter to show a loss and thereby earn more guarantee.

TABLE 19

ENGLISH CENTRAL MILLS IN PERNAMBUCO

Harvest/ Performance	Firmeza (Escada)	Santo Ignácio (Cabo)	Cuyambuca (Água Preta)	Bom Gôsto (Palmares)	Tiúma (São Lourenço da Mata)
1884-85					
Cane Milled (tons)	6,393	12,228	6,638	8,923	-
Syrup/Cane (percent)	60.7	56.0	60.5	56.0	-
Sugar Made (tons)	560	677	547	568	-
Sugar/Cane (percent)	8.8	5.5	8.3	6.4	-
Expenses (contos)	141	213	135	179	-
Receipts (contos)	77	99	77	83	-
Deficit (contos)	64	114	58	96	-
1885-86					
Cane Milled	11,215	14,316	11,198	9,781	-
Syrup/Cane	61.5	61.5	66.2	59.5	-

Sugar Made	789	728	963	495
Sugar/Cane	7.0	5.1	8.1	5.1
Expenses	212	246	215	187
Receipts	166	157	188	104
Deficit	46	89	27	83
1887-88				
Cane Milled	–	–	–	32,205
Syrup/Cane	–	–	–	60.7
Sugar Made	–	–	–	1,837
Sugar/Cane	–	–	–	8.0
Expenses	–	–	–	294
Receipts	–	–	–	252
Deficit	–	–	–	42

SOURCES: "Relatório apresentado ao Ministério dos Negócios da Agricultura, Commercio e Obras Publicas pelo Engenheiro Fiscal Francisco do Rego Barros," *Annexos ao Relatório apresentado à Assembléa Geral na Segunda Sessão da Vigésima Legislatura pelo Ministro e Secretário de Estado dos Negócios da Agricultura, Commercio e Obras Publicas Rodrigo Augusto da Silva* (Rio de Janeiro, 1887), Appendix A, tables following p. 17. "Relatório . . . Engenheiro Fiscal Francisco do Rego Barros," 1889, p. 4.

97

referred to the company's "poor administration," "the inexpert-
ness of their technical directors, the squanderings of management and the London directors, and the lack of capable workers."[22] In a vain attempt to control the situation the imperial government named an Inspector Engineer in 1884 and an Inspection
Committee in 1885, but their powers were too limited.[23]

The company itself attributed its failure to insufficient quantities of poor-quality, high-cost cane; to low sugar prices; to
mechanical problems; and to interfering Brazilian officials. Certainly sugar prices fell violently between 1882 and 1886 (Table
4). But the deficits provoked were covered by the imperial government: the CSFB received 644 contos or 7.7 percent annual
return over three harvests on a nominal investment of 2,800
contos. The inferior materials and machinery, and incompetent
management, must be blamed on the company itself. In 1886
the CSFB began liquidation.[24]

Groups of former CSFB cane suppliers managed to salvage
something from the company's failure. They rented the Firmeza,
Cuyambuca, and Bom Gôsto mills from the company for 1 percent
of the sugar made and maintenance. Only the Cabo mill eventually fulfilled its original purpose, however; in the 1890's, refur-

22. *Falla com que o Excm⁰ Conselheiro José Fernandes da Costa Pereira Junior abriu a sessão da Assembléa Legislativa da Provincia de Pernambuco no 6 de março de 1886*, pp. 50-51. "Relatório. . . Engenheiro Fiscal Francisco do Rego Barros," 1886, p. 6. Raffard, *O Centro da Indústria*, p. 116.
23. Francisco do Rego Barros to Provincial President, January 22, 1886, September 16, 1885, both in APE, Coleção Engenhos Centraes. Antonio Augusto Fernandes Pinheiro to provincial president, October 8, 1885, *ibid.*
24. "Relatório . . . Engenheiro Fiscal Francisco do Rego Barros," 1887, pp. 68. Atherton to U.S. Department of State, Recife, October 18, 1886, cited in Graham, *Britain and the Onset of Modernization*, p. 155. "Engenhos Centraes no Brazil," *Diário de Pernambuco*, November 30, 1886. Joaquim Corrêa de Araujo, "Publicações a pedido. A Companhia Central Sugar Factories of Brazil, Limited," *ibid.*, February 18, 1887. Francisco do Rego Barros to provincial president, September 6, 1888, APE, Coleção Engenhos Centraes. *Officio com que o Excm⁰ Sr. Presidente Conselheiro José Fernandes da Costa Pereira Junior entregou a administração da provincia de Pernambuco ao Excm⁰ Sr. 1⁰ Vice-presidente Dr. Ignacio Joaquim de Souza Leao em 30 de março, e Relatório que o mesmo Excm⁰ Sr. Vice-presidente apresentou ao Excm⁰ Sr. Presidente Dr. Pedro Vicente de Azevedo em 10 de novembro de 1886*, p. 56.

bished with state government aid, it became the successful Usina Santo Ignácio.[25]

The sorry CSFB episode was not an isolated incident. The North Brazilian Sugar Factories, Limited (NBSF), founded by an English railroad-building concern and a group of London civil engineers, behaved similarly. Presumably in exchange for stock, the NBSF in 1882 acquired from two Brazilians two central mill concessions guaranteeing 6 percent on a total of 3,750 contos to build seven mills. But by 1886 no NBSF mill had begun grinding. The delay resulted from false cane-supply contracts by one of the Brazilian concessionaires, the business failure of an import-export house illegally subcontracted by the company to buy machinery, and the company's attempt to finance construction from advances on its guarantee—in other words, to build with the government's money.[26]

Disgusted with the foreigners' procrastination, the imperial

25. "Relatório apresentado ao Ministério dos Negocios da Agricultura, Commercio e Obras Publicas pelo Engenheiro Fiscal Francisco do Rego Barros," *Annexos ao Relatório apresentado à Assembléa Geral na Terceira Sessão da Vigesima Legislatura pelo Ministro e Secretario de Estado dos Negocios da Agricultura, Commércio e Obras Publicas Rodrigo Augusto da Silva* (Rio de Janeiro, 1888), Appendix C, p. 3. Carlos Beltrão, "Engenhos Centraes," *Diário de Pernambuco,* July 29, 1888. Recife and San Francisco Pernambuco Railway Company (Limited), *Report of the Proceedings at the Sixty-Fourth Half-Yearly Ordinary General Meeting of the Shareholders . . . October 11, 1887,* p. 1.

26. SNA. *Legislação Agrícola,* v. II, Part 2, pp. 265-300, for original NBSF statutes and the concession. The railway company was Reed Bowen and Company, which had begun construction on the Natal-Nova Cruz line in July 1871 and opened it to service in September 1881. Francisco Picanço, *Diccionario de Estradas de Ferro,* 2 vols. (Rio de Janeiro, 1891-92), v. II, pp. 87, 89. "North Brazilian Sugar Factories, Limited," *Diário de Pernambuco,* November 29, 1882. The Brazilian concessionaires were Domingos Moitinho and Jovino Bandeira. The company had received 373 contos in advances by 1886, without having made any sugar. "Relatório . . . Engenheiro Fiscal Francisco do Rego Barros," 1886, p. 5. "Relatório . . . Engenheiro Fiscal Francisco do Rego Barros," 1888, p. 4. *Relatório apresentado à Assembléa Geral na Segunda Sessão da Vigesima Legislatura pelo Ministro e Secretario de Estado dos Negocios da Agricultura, Commercio e Obras Publicas Rodrigo Augusto da Silva* (Rio de Janeiro, 1887), p. 16. "Relatório . . . Engenheiro Francisco do Rego Barros," 1887, pp. 12-13. Francisco do Rego Barros to the provincial president, January 22, 1886, APE, Coleção Engenhos Centraes.

government in later 1886 cancelled all but three Pernambuco concessions and suspended the guaranteed return on two. The company responded by reorganizing and reducing its capital by over two-thirds. In 1887 the company finally put the Tiúma mill at São Lourenço da Mata into operation. Although initial results far exceeded the best CSFB performance, the company ran a deficit nevertheless (Table 19). Blaming high cane prices, despite the fact that cane-supply contracts included two-way escalator clauses relating cane prices to sugar prices, the company curtailed loans to cane growers, obliged them to pay one-half the cane transport cost, and improved efficiency. The NBSF refused to build adequate waste dumps to prevent pollution of the river, which killed fish basic to the diet of the townspeople; and it sent home the English manager who, like his unhappy counterpart at the CSFB's Bom Gôsto mill, had become an intemperate consumer of distillery products.[27]

The Tiúma mill prospered, and by 1895 the national government could suspend its interest guarantee. Again Brazilian planters salvaged material from the wreckage of other NBSF concessions. João Zeferino Pires de Lyra bought the machinery from the

27. "Indice Geral e Chronologico das Leis e Decretos . . . até o Fim do Anno de 1887," p. 96. The other sites were at Nazaré and Pau d'Alho. SNA, *Legislação Agrícola*, v. II, Part 2, pp. 299, 350. "Relatório apresentado ao Ministerio dos Negócios da Agricultura, Commercio e Obras Publicas pelo Engenheiro Fiscal Francisco do Rego Barros," *Annexos ao Relatório apresentado à Assembléa Geral na Quarta Sessão da Vigesima Legislatura pelo Ministro e Secretario de Estado Interino dos Negocios da Agricultura, Commercio e Obras Publicas Rodrigo Augusto da Silva* (Rio de Janeiro, 1889), v. I, p. 4. Francisco do Rego Barros to provincial president, February 2, June 27, October 29, November 6, 1888, APE, Coleção Engenhos Centraes. The townspeople became so incensed at the company's "impudent" indifference to pollution that the provincial president had to send "a strong detachment of troops to contain them." *Relatório com que o Exm. Sr. Desembargador Joaquim José de Oliveira Andrade entregou a administração da Província ao Exm. Sr. Dr. Innocencio Marques de Araujo Goes em 3 de janeiro de 1889*, p. 37. J. J. Adam to provincial president, September 27, 1889, APE, Coleção Engenhos Centraes. "Relatório . . . Engenheiro Fiscal Francisco do Rego Barros," 1889, p. 5. The provincial president agreed with the imperial Inspector-Engineer that "poor administration" prevented the mill's success. *Falla que à Assembléa Legislativa Provincial de Pernambuco no dia de sua installação a 15 de setembro de 1888 dirigio o Exm. Sr. Presidente da Provincia Desembargador Joaquim José de Oliveira Andrade*, p. 38.

Pau D'Alho concession in 1887 and built his 13 de Maio mill. Manoel Antonio dos Santos Dias bought machinery from an unsuccessful NBSF mill in Rio Grande do Norte for his Engenho Jundiá, which he later converted into the Usina Santa Philonila.[28]

The careers of the CSFB and the NBSF in Pernambuco hardly justify concluding that "the British-owned central sugar factories made substantial contributions toward the modification of the northeast," and that their failures resulted from "their position as innovators inadequately dealing with a hostile environment."[29] Nor can one affirm that "the appearance of *engenhos centrais* (central sugar mills) and *usinas* (big mills), was appreciably weakening the old regime in a manner suggestive of the Cuban experience but on a much smaller scale."[30] A French traveler in the late 1880's came closer to the truth when he castigated the English, "who know how to get such beautiful results in Demerara, their Guayana colony, [but] have done nothing in Brazil but pump out the subsidies, the interest guarantees, without performing any real service of even a little importance for agriculture."[31]

In contrast to the dismal English record, Brazilians mounted

28. *Relatório apresentado ao Presidente da Republica dos Estados Unidos do Brazil pelo Ministro de Estado dos Negocios da Industria, Viação e Obras Publicas Sebastião Eurico Gonçalves de Lacerda em Maio de 1898, 10° da Republica*, p. 18. "Indice Geral e Chronologico das Leis e Decretos . . . até o Fim do Anno de 1887," p. 99.

29. Graham, *Britain and the Onset of Modernization*, pp. 156, 149. Graham himself recognizes the emptiness of this claim when he bases it not on English performance but on what Brazilian planters later did with the companies' material. Curiously, Graham cites "their original and misguided concentration on the export market" as one of the companies' liabilities. But the English concentration was inevitable: as late as the early twentieth century Brazil's domestic market still was unable to absorb all Pernambuco's sugar at a remunerative price.

30. Eugene D. Genovese, *The World the Slaveholders Made. Two Essays in Interpretation* (New York, 1969), p. 91. For a more precise definition of the two types of mill, see below.

31. Marc, *Le Brésil*, v. I, pp. 261-262. For critical English views, see Wells, *Exploring and Travelling Three Thousand Miles Through Brazil*, v. II, pp. 344-346, and Arthur L.G. Williams, "Report for the Year 1894 on the Trade of the Consular District of Pernambuco," *Parliamentary Papers*, 1895, HCC, v. 96, *AP*, v. 36, p. 16.

five successful central mills in Pernambuco. All five enjoyed provincial government subsidies whose terms reflected lessons learned from the imperial government's experience. Provincial Law 1,860 of August 11, 1885, sanctioned direct loans rather than guaranteed returns. The province would loan a total of 800 contos in 6 percent and 7 percent bonds, which the concessionaire would sell locally against mortgages on the properties, while the concessionaire could only transfer his rights to planters. Laws 1,971 and 1,972 of March 2, 1889, increased the total loans available to 1,700 contos, required that concessionaires be planters residing in the county of the planned mill, and forbid them to transfer the concession before beginning construction. The concessionaires had to post the value of their subsidy in securities or real estate until the mortgages were registered, and the province would pay the first half of the subsidy only after mill plans were approved, and the second half after construction and installations were finished.[32]

Between 1885 and 1890, the provincial president received more than thirty different proposals to build central mills under Laws 1,860, 1,971, and 1,972. Five concessionaires, each subsidized with 200 contos, built mills, and five of these were successful. José da Silva Loyo Júnior transferred his concession to the Companhia Uzina João Alfredo, which began grinding cane in Goiana in 1889. The company reorganized as the Companhia Industrial Pernambucana and repaid its debt by 1891. João Zeferino Pires de Lyra used his subsidy to buy NBSF machinery and built the central mill 13 de Maio in Palmares. Gaspar de Menezes Drummond built the central mill Trapiche in Sirinhaém, and Colonel João Carlos de Mendonça Vasconcellos and Captain João Paulo Moreira Temporal located their central mill Carassú in Barreiros.[33] The mill in Barreiros was considered the "best in the

32. "Lei 1,860," articles 16-28, Collecão de Leis Provinciaes de Pernambuco, Anno de 1885, pp. 46-49. "Lei 1,971," "Lei 1,972," Collecão de Leis Provinciais da Provincia de Pernambuco publicada no anno de 1889, pp. 6-8.
33. Ibid. Officio . . . José Fernandes da Costa Pereira Junior . . . 30 de março . . . Relatório . . . (Ignacio Joaquim de Souza Leão) . . . 10 de Novembro de 1886, p. 56. Portaria by provincial vice-president, June 23, 1886, APE, Colecão Portarias, v. 66. Relatório com que o Exm. 1º vice-presidente Dr. Ignacio Joaquim de Souza Leão passou a administração da provincia em 16 de

province."[34] Leocadio Alves Pontual and Joaquim Ignacio Pessoa de Siqueira transformed their Engenho Aripibú in Amarají with their subsidy. Finally, Antonio Carlos de Arruda Beltrão built the sugar refinery Usina Beltrão between Olinda and Recife. Acquired by the Companhia Industrial Assucareira in Rio de Janeiro, the refinery began operations in 1894; while it was reportedly the "largest in Brazil" and its French machinery was "of the very best make," by 1899 it had "proved a complete failure" for lack of capital.[35] The five successes can be attributed to the form

abril de 1888 ao Exm. Presidente Desembargador Joaquim José de Oliveira An- drade, p. 18. "Engenhos Centraes para Pernambuco," *Diário de Pernambuco,* April 25, 1888. "Engenho Central," *ibid.*, June 23, 27, 1888, May 21, 24, 25, 1889, *Relatório com que o Exm. Sr. Desembargador Joaquim José de Oliveira Andrade entregou a administração da provincia ao Exm. Sr. Dr. Innocencio Marques de Araujo Goes em 3 de janeiro de 1889,* pp. 34-36. Francisco Apoligonio Leal to provincial president, October 20, 1887, APE, Coleção Obras Publicas, v. 63. Various letters to provincial president, April 8, May 19, 1889, APE, Coleção Engenhos Centraes. "Sociedade anonyma. Uzina João Alfredo," *Diário de Pernambuco,* December 1, 1888. "Cardozo e Irmão ao respeitável publico," *ibid.*, May 26, 1889. *Mensagem apresentada ao Congresso Legislativo do Estado, em 10 de agosto de 1891, pelo vice-governador Desembargador José Antonio Corrêa da Silva,* pp. 49-50. Gonçalves e Silva, *O Assucar e o Algodão em Pernambuco* (Recife, 1929), p. 54, Peres and Peres, *A Industria Assucareira em Pernambuco,* p.146. "A Provincia. Engenhos centraes," *A Provincia,* June 26, 1888. João Zeferino Pires de Lyra, "Publicações a Pedido. Engenho central 13 de Maio," *Diário de Pernambuco,* August 9, 1890.

34. "Engenhos Centraes," *ibid.*, November 19, 1887. Gaspar Menezes Drummond to provincial president, January 5, 17, 1889; Francisco de Souza Reis to Provincial President, November 13, 28, 1889, all in APE, Coleção Engenhos Centraes. *Mensagem. . . 10 de agosto de 1891. . . José Antonio Corrêa da Silva,* p. 49.

35. Paulo de Amorim Salgado and others to provincial president, June 6, 1890, APE, Coleção Diversos, v. 44. "Contracto," June 16, 1890, APE, Coleção Portarias, v. 78. Manoel Nicolau to governor, August 17, 1891, APE, Coleção Tesouro do Estado, v. 11. *Relatório com que o Desembargador Barão de Lucena entregou a 23 de Outubro de 1890, o governo do Estado de Pernambuco ao Desembargador José Antonio Corrêa da Silva,* pp. 36-37. "Contracto," November 2, 1889, APE, Coleção Engenhos Centraes. *Mensagem . . . 10 de agosto de 1891 . . . José Antonio Corrêa da Silva,* p. 49. "Contracto," October 31, 1891, APE, Coleção Tesouro do Estado, v. 12. "Usina Beltrão," *Diário de Pernambuco,* February 2, 1895. Consul Howard, "Report on the Trade and Commerce of the Consular District of Pernambuco, for the Years 1899-1900," *Parliamentary Papers,* 1901, HCC, v. LXXXI, *AP,* v. 45, p. 9. Arthur L. G. Williams, "Report for the Year 1894 on the Trade of the Consular District of Pernambuco," *ibid.*, 1895, HCC, v. 96, *AP,* v. 36, p. 9. Arruda Beltrão, *A Lavoura da Canna,* p. 21.

of the subsidy, a direct loan, and to the nature of the recipients, Brazilian planters interested in making sugar, not in speculation.

The imperial government made its last attempts to promote central mills in the eighteen months following the abolition of slavery, and undoubtedly as indirect compensation for sugar interests. Decree 10,100 of December 1, 1888, and Decree 10,393 of October 9, 1889, raised individual subsidies available under previous imperial legislation to allow larger subsidies for smaller mills, gave preference to companies using diffusion, set cane prices at one twenty-fifth of mascavado sugar prices, and required 7 percent sugar-to-cane in the first harvest. Pernambuco received a slightly reduced quota, but it was still the largest in the Empire (Table 18). The first Republican government doubled the capital available by means of its own Decree 525. Although at least one dozen concessions were awarded in Pernambuco under these three laws, no central mills resulted: four were canceled by 1891, four were transferred but not used, and four others were made and never again mentioned.[36]

36. "Decreto 10,100," *Colleção de Leis do Império do Brazil de 1888*, parte II, Tomo LI, v. II, pp. 466-474. "Decreto 10,393," *Colleção de Leis do Imperio do Brazil de 1889*, parte II, tomo LII, v. II, pp. 481-489. Under Laws 2,687 and 8,357, to earn 7 percent on 1,000 contos a mill had to grind 80,000 tons of cane per harvest. Decrees 10,100 and 10,393 lowered the guaranteed return to 6 percent but in compensation required that the mill grind only 50,000 tons of cane, about a 40 percent reduction. Moreover, if the concessionaire established immigrant colonies he did not have to present cane supply contracts, no deposit was required, and if unsuccessful, the company had no obligation to reimburse the government for advances received. José Honório Rodrigues, who also glossed this central mill legislation, mistakenly wrote millions and hundreds of thousands of contos where he meant thousands and hundreds of contos. "A Revolução Industrial Açucareira, I," p. 182; *ibid.*, II, p. 229. "Decreto 525," *Decretos do Governo Provisorio da Republica dos Estados Unidos do Brazil, Sexto Fasciculo de 1 a 30 de junho de 1890*, pp. 1,421-422. *Relatório apresentado ao Presidente da Republica dos Estados Unidos do Brazil pelo Ministro d'Estado dos Negocios da Agricultura,Commercio, e Obras Publicas Barão de Lucena em junho de 1891*, pp. 8-9. *Relatório apresentado ao Vice-Presidente da Republica dos Estados Unidos do Brazil pelo Ministro d'Estado dos Negocios da Agricultura, Commercio, e Obras Publicas Engenheiro Antão Gonçalves de Faria em Maio de 1892*, p. 9. *Relatório apresentado ao Vice-Presidente da Republica dos Estados Unidos do Brazil pelo Ministro de Estado dos Negocios da Industria, Viação e Obras Publicas Engenheiro Antonio Francisco de Paula Souza no Anno de 1893, 5° da Republica*, p. 22. The Agriculture Ministry was

The sterility of Laws 10,100, 10,393, and 525 ended the era of central mills in Pernambuco. On balance, the results had not been impressive. Those three laws and their imperial and provincial predecessors had envisioned a minimum of 31 central mills and a total of 20,500 contos. Only six mills were actually built with subsidies of about 2,800 contos. Central mill subsidies failed in Pernambuco for two reasons. First, speculation, fraud, and bad faith obviously had most to do with the failures. Even the regulations legislated to control such behavior at times became so severe as to be counterproductive: a state official criticized the regulations for Provincial Laws 1,860, 1,971, and 1,972 by noting that anyone who could put up collateral equal to the subsidy or complete construction before receiving the full grant "certainly would not seek the aid of the state with so many obligations and requirements."[37]

Secondly, those mills that operated a few years and then stopped, or that ran long-run deficits, probably suffered from lack of control over cane supply and prices. Many planters learned from the unhappy experiences of the earliest suppliers to CSFB mills, and they conserved their engenho machinery and thereby their independence from the central mill. When they did not like the cane price, they simply ground the cane themselves. Modern

abolished in 1892, and this latter ministry assumed its responsibilities. *Relatório apresentado ao Vice-Presidente da Republica dos Estados Unidos do Brazil pelo General de Brigada Dr. Bibiano Sergio Macedo da Fontoura Costallat Ministro de Estado dos Negocios da Industria, Viação e Obras Publicas em Maio de 1894,6° da Republica*, pp. 24-25. SNA, *Legislação Agricola*, v. II, Part 2, pp. 364-366, 373. *Relatório apresentado ao Chefe do Governo Provisorio por Francisco Glicerio Ministro e Secretário de Estado dos Negocios da Agricultura, Commercio e Obras Publicas* (Brazil, 1890), pp. 54-55.

37. To estimate the minimum number of mills planned, we have divided the total amount of money available by the largest individual subsidy permitted. To estimate the total subsidy to the NBSF Tiúma mill, we prorate the 373 contos paid between 1882 and 1886 until 1895, and round off. Probably this estimate of 1,000 contos exaggerates, for Tiúma made profits after the first few harvests. Eduardo Augusto d'Oliveira to governor, June 9, 1890, APE, Coleção Tesouro do Estado, v. 3. Oliveira was criticizing the regulation of September 18, 1889, which was more stringent than Laws 1,971 or 1,972: amortization had to begin after the first harvest, and the concessionaire had to pay the Inspector-Engineer's salary until amortization was complete, rather than simply until construction ended.

Brazilian historians have emphasized this factor. Paul Singer affirms that "the fundamental cause for the failure of the central mill is that it could not count on a sure supply of cane at prices which allowed it to make sugar at a competitive cost." Miguel Costa Filho basically agrees when he explains that "the central mills tended to benefit a large number of producers [the cane growers], instead of only one," and thereby inhibited capital accumulation and lowered the profit rate. Gileno Dé Carli identifies "the instability of raw material supply" as the principal cause.[38]

Usinas

The failure of central mills in Pernambuco did not discourage interest in modernizing the sugar industry. The first decade of the Republic witnessed a happier event—the birth of the usinas. The usina differed from the central mill principally in regard to the division of labor. While the central mill had specialized in sugar manufacture and bought its cane supply, the usina not only bought cane from planters, known as *fornecedores,* but also grew its own supply. Initially this cane grew on the plantation where the usina was built. But to keep the large mill running efficiently the usineiro had to insure a steady supply of cane, and he did so by gradually acquiring the fornecedores' plantations. Thus the usina represented the third stage of a dialectical process. From the sixteenth until the later nineteenth century, the traditional engenho made sugar from cane grown almost exclusively on its own land. Then the contradictions of market demand and Brazilian supply required improving the product and lowering costs. The attempted solution through specialization—establishing central mills—failed, and almost immediately the essential dependency of mill and cane field led to a new vertical integration,

38. Arthur L.G. Williams, "Report for the Year 1894 on the Trade of the Consular District of Pernambuco, *Parliamentary Papers,* 1895, HCC, v. 96, *AP,* v. 36, pp. 15-16. Singer, *Desenvolvimento Econômico e Evolução Urbana,* p. 297. Miguel Costa Filho, "Engenhos Centrais e Usinas," *Revista do Livro* (Rio de Janeiro), Ano V, no. 19 (September 1960), p. 88. Gileno Dé Carli, *O Processo Histórico da Usina em Pernambuco* (Rio de Janeiro, 1942), p. 10. Alice Canabrava, "A Grande Lavoura," p. 109, also uses this explanation.

accomplished by having the usina own canefields and gradually absorb independent cane plantations. In effect, the usina was a modern reincarnation of the traditional engenho on a more complicated and far larger scale.[39]

The early Republican governments initiated official support for usinas with impressive success. The Barão de Lucena, who had supported central mills when he served as provincial president in the 1870's, approved the first law on October 15, 1890. This law, revoking Provincial Laws 1,860, 1,971, and 1,972, offered loans of 200 contos in 7 percent state bonds to planters building "small usinas" producing 900 tons of sugar per harvest. To insure a cane supply, the law tied cane prices to sugar prices and established a zoning principle: "no usina will be granted when its proximity to others already built or to be built with or without subsidies could compromise the future of the new enterprise by the lack or scarcity of cane supply." The state would hold the first mortgage on lands and the factory; it would pay the first installment of the loan when materials and machinery were purchased, the second when equipment was on site, and the third when construction and installation were finished. A final installment would be paid after the first harvest. If the usineiro failed to pay interest or 5 percent amortization for two years, he would incur a 1 percent monthly penalty interest and the state would assume management of the usina to pay itself from revenues earned. To compensate for exchange rate falls, Lucena's successor José Antonio Corrêa da Silva raised the subsidy to 250 contos; he also required that concessionaires themselves supply at least 50 percent of the cane.[40]

In the next few years the state government subsidized many usinas under these laws. Lucena and Corrêa da Silva granted

39. Except for the dialectics, this analysis also appears in Dé Carli, *O Processo Histórico da Usina*, pp. 12-20. *Idem., Aspectos da Economia Açucareira* (Rio de Janeiro, 1942), pp. 17-18. Barbosa Lima Sobrinho, *Problemas Econô-micos e Sociais da Lavoura Canavieira* (Rio de Janeiro, 1943), pp. 21-23. Miguel Costa Filho, "Engenhos Centrais e Usinas," pp. 89-91.

40. *Colleção de Leis Estaduais de Pernambuco, Anno de 1890*. Paulo Cavalcanti kindly gave me a copy of the October 15 law. "Parte Official," *Diário de Pernambuco*, December 13, 1891. "Ainda as usinas," *ibid.*, July 1, 1891.

TABLE 20

PERNAMBUCO CENTRAL MILLS AND USINAS TO 1910

First Harvest[a]	Name[b]	Location[c]	Subsidy Amount (contos)	Date[d]
1874	São Francisco da Várzea	Várzea (Recife)		
1875	São José da Várzea	Várzea (Recife)		
1877	Mameluco	Escada		
1878	Tinoco	Sirinhaém		
1881	Massauassú	Escada	600	1895
1881	Limoeirinho	Escada		
1884	Santo Ignácio	Cabo	7% on 700	1881
	Santo Ignácio	Cabo	600	1895
	Firmeza	Escada	7% on 700	1881
	Cuyambuca	Água Preta	7% on 700	1881
	Bom Gôsto	Palmares	7% on 700	1881
1885	Nova Conceição	Ipojuca	250	1895
1886	Colônia Isabel	Palmares		
1887	Frei Caneca	Palmares (Maraial)		
	Timbó	Olinda (Paulista)		
	Tiúma	São Lourenço da Mata	6% on 750	1882

Year	Engenho	Município	Capacity	Year
	Bom Destino (Treze de Maio)	Palmares	200	1888[e]
	Pinto (Pão Sangue, Santa Cruz-Ribeirão)	Gameleira (Ribeirão)	500	1895
			800	1895
			800	1896
1888	Bandeira (Ipojuca)	Ipojuca	f	1890[f]
	Aripibú	Escada (Amarají)	200	1895[f]
	Cabeça de Negro	Escada (Amarají)	250	1889[f]
	Carassú	Barreiros	200	1891[f]
	Bamburral	Escada (Amarají)	250	1895
1889	Muribeca	Jaboatão (Muribeca)	550	1888
	João Alfredo	Goiana	200	1891[e][f]
	N.S. das Maravilhas	Goiana	200	1887
	Santa Philomila	Escada	200	1891
1890	Trapiche	Sirinhaém	250	1895
1891	Estrelliana	Gameleira	750	1891
	Pirangy	Palmares	250	1895
	Maria das Mercês	Cabo	250	1890
	Maria das Mercês	Cabo	100	1891
	Guerra	Ipojuca	200	1891
	Guerra	Ipojuca	250	1891
?	Cursahy	Pau d'Alho	250	1891
?	Ilha das Flôres	Gameleira	250	1891
?	Lustosa (Phenix)	Garanhuns (Quipapá)	250	1891

(Continued)

First Harvest	Name	Location	Subsidy Amount (contos)	Date
?	Lustosa (Phenix)	Garanhuns	350	1895
1892	Correia da Silva (Catende)	Palmares (Catende)	250	1891
	Florestal	Palmares (Catende)	800	1895
	Pedrosa	Garanhuns (Quipapá)	250	1891
	Roçadinho	Bonito		
	Salgado	Bonito	250	1891
	Salgado	Ipojuca	100	1895
	Cachoeira Lisa	Ipojuca	250	1891 [c]
1894	São João	Gameleira		
1895	Bosque	Recife	[f]	
	Bom Jesus	Escada		
	Bulhões	Cabo		
	Caxangá	Jaboatão		
	Bom Fim (União e Indústria)	Gameleira (Ribeirão)	600	1895 [h]
	Pery-Pery	Escada	600	1895
?	Beltrão (Tacaruna)	Garanhuns (Quipapá)	200	1889 [f,i]
?	Barão de Morenos	Recife	600	1895 [j]
?	Conceição	Jaboatão	250	1895
?	Espírito Santo	Vitória	800	1895 [j]
	N.S. de Lourdes	Pau d'Alho		
		Jaboatão	600	1895

?	Raíz de Dentro	Amarají	200	1895
?	São José	?	800	1895 [j]
1896	Jaboatão	Jaboatão		
	Cucaú	Rio Formoso		
1897	Timbó Assú	Escada		
1900	Frexeiras	Escada	600	1895
	Mussú	Escada		
1905	Meio da Várzea	Recife		
1906	Mussupe	Igarassú	250	1891
	(Coelho, São José)	Igarassú	350	1895
	Ubaquinha	Sirinhaém		
1907	Destêrro	Pau d'Alho		
1910	Santa Thereza	Jaboatão	500	1895
	(Progresso Colonial)			

[a] Where first harvest unknown, we use a question mark (?); several of these mills never functioned.
[b] Where mill renamed, later name or names in parentheses.
[c] Where location later separated from original county, new county in parentheses.
[d] Only date of first contract, not revised contracts.
[e] Equipped initially with discarded North Brazilian Sugar Factories Limited machinery, an indirect subsidy.
[f] Also received Banco de Crédito Real loan.
[g] Also received 120 contos to build a railroad.
[h] Also received 720 contos to build a railroad.
[i] An urban refinery; it did not buy cane.
[j] Subsidy later canceled.

(*Continued*)

SOURCES: APE, Coleção Engenhos Centrais, passim. Mensagem . . . 10 de agôsto de 1891 . . . José Antonio Corrêa da Silva, passim.

Mensagem apresentado pelo Governador do Estado de Pernambuco Conselheiro Joaquim Corrêa d'Araujo ao Congresso Legislativo do Estado em 27 de outubro de 1896, passim.

Mensagem . . . 6 de março de 1899 . . . Joaquim Corrêa de Araujo, passim.

Relatório apresentado ao Exmo. Sr. Antonio Gonçalves Ferreira, Governador do Estado, pelo Secretário da Industria, Obras Publicas, Agricultura, Commercio e Hygiene (Recife, 1902), passim. This Relatório transcribes nearly all contracts made by state governments.

"Secretário da Industria, Relatório apresentado ao Exmo. Sr. Dr. Júlio de Mello Filho, Secretário do Estado do Negócios da Industria, pelo Bacharel João Diniz Ribeiro da Cunha, Director Geral da 1° Diretoria," Diário de Pernambuco, July 4, 1896.

Gonçalves e Silva, O Assucar e o Algodão, passim.

Peres and Cavalcanti, Industrias de Pernambuco, passim.

Veríssimo de Toledo, Almanack Administrativo, Mercantil, Agrícola e Industrial do Estado de Pernambuco, passim.

Julio Pires Ferreira, Almanack de Pernambuco para o anno de 1900, passim.

fourteen concessions totaling 3,400 contos in bonds: three of these concessions aided mills already in operation, and the other eleven loans led to the construction of at least six new usinas by 1896 (Table 20). These successes are all the more impressive since between 1890 and 1896 the value of the mil-réis dropped from 14.9d to 7.7d; in other words, machinery imports became nearly twice as expensive.[41]

Alexandre José Barbosa Lima, the third governor to subsidize usinas in Pernambuco, has been justly remembered as the principal state usina promoter in the Republic. His reputation derives from both the number and the value of his concessions. From the time of his first loan in December 1894 "until the last days of his government, more or less, his Excellency gave large bond loans to everyone who asked."[42] Barbosa Lima's concessions were nearly all made under the terms of State Law 113 of June 22, 1895. This law amplified the 1890 and 1891 usina laws by raising the base loan to 500 contos. For every additional 375 tons of sugar produced the mill could receive an additional 100 contos, as well as 8 contos additional for every kilometer of railroad in excess of ten kilometers. The law also tied the loan to the exchange rate as an inflation hedge. Barbosa Lima made 23 loans totaling 11,450 contos, over triple the amount loaned by Lucena and Corrêa da Silva; the average loan was 500 contos, and five usinas received 800 contos each. Seven loans helped start new usinas; the other sixteen enabled usineiros to complete or expand usi-

41. *Mensagem . . . 10 de agosto de 1891 . . . José Antonio Corrêa da Silva*, pp. 48-50. The governor listed thirteen concessions, but he omitted the one made to build the Usina Maria das Mercês in Cabo. "Parte official," *Diário de Pernambuco*, July 26, 1891. For dates of first harvests, see Gonçalves e Silva, *O Assucar e o Algodão*. Barbosa Lima Sobrinho also notes the exchange fall, and concludes that it nullified the effects of the 1890 and 1891 loans. But he is only correct for the five fruitless concessions. Barbosa Lima Sobrinho, "O Governo Barbosa Lima e a Industria Açucareira de Pernambuco," *Annuario Açucareiro para 1938*, p. 363.

42. Barbosa Lima Sobrinho, "O Governo Barbosa Lima, p. 364. The praise comes not only from his nephew, Barbosa Lima Sobrinho, *Problemas Econômicos e Sociais da Lavoura Canavieira*, p. 21, but also from Gileno Dé Carli, *O Processo Histórico da Usina*, pp. 13-14, and Paul Singer, *Desenvolvimento Econômico e Evoluçao Urbana*, p. 299.

nas already begun. Only three concessions were canceled, and only two of the loans made were never used.[43]

Barbosa Lima's loans constituted the largest and also the last attempt to promote usinas in Pernambuco with direct subsidies. By the end of his term the usina bonds had "flooded" Recife's securities market and "their value almost immediately dropped from par [1 conto] to 750 mil-réis and 700 mil-réis each." According to the British consuls, the bonds were discounted heavily because of their excessive quantity, not because buyers regarded state-supported usinas as a dubious investment.[44]

Instead of new loans, the governors after Barbosa Lima encouraged those usineiros already in debt to the state treasury

43. Barbosa Lima Sobrinho, "O Governo Barbosa Lima," pp. 364-367. *Mensagens apresentadas ao Congresso Legislativo do Estado em 1893, 1895, e 1896 pelo Dr. Alexandre José Barbosa Lima quando governador de Pernambuco* (Recife, 1931), pp. 231-236. In return for the larger loans, the concessionaires had to permit public use of their railroads, reforest cut-down areas, and publish all documents relative to their concession. Barbosa Lima Sobrinho gives a total of 13,950 contos, but without deducting the cancelled concessions to the Usinas Espirito Santo, Barão de Morenos, and São José. Governor Barbosa Lima himself gave a total in excess of 12,000 contos, but he was referring to all state loans to usinas, not only those during his regime. Alexandre José Barbosa Lima, *Discursos Parlamentares*, 2 vols. (Brasilia, 1963-69), v. I, p. 374. For cancellation notices, see *Mensagem . . . Joaquim Corrêa d'Araujo . . . 27 de outubro de 1896*, pp. 38-39. In 1905, a State Treasury official described five projected usinas as still unbuilt: Nossa Senhora de Lourdes, Pery-Pery, Raiz de Dentro, Pão Sangue, and Santa Cruz. But the Companhia Geral de Melhoramentos eventually used the last two concessions to expand the successful Usina Ribeirão, and Pery-Pery, under different ownership, later reached completion. The Treasury official also characterized the Usinas Coelho, Conceição, Nova Conceição, Progresso Colonial, Muribeca, 13 de Maio, and Santo Ignacio as abandoned and stripped of movable parts. But the Usinas Muribeca, 13 de Maio, and Santo Ignacio were milling again by 1929. The other mills may have remained in *fogo morto*, or may have resumed operations under a different name. "Relatório apresentado pelo Escrivão da despeza do Thesouro José de Goes Cavalcanti, sobre a escripturação das Usinas auxiliadas pelo Estado, cujos productos se acham sujeitos aos impostos da lei orçamentaria vigente—art. 1⁰, par. 2," *Diário de Pernambuco*, September, 1, 3, 5, 1905. Gonçalves e Silva, *O Assucar e o Algodão.*

44. A.F. Howard, "Report for the year 1896 on the Trade, etc., of the Consular District of Pernambuco," *Parliamentary Papers*, 1897, HCC, v. 89, AP, v. 28, pp. 12-13. Vice-Consul Williams, "Report on the Trade and Commerce of the Consular District of Pernambuco for the Year 1897," *ibid.*, 1898, HCC, v. 94, AP, v. 43, p. 3.

by facilitating repayment. Governor Joaquim Corrêa de Araujo criticized Barbosa Lima for committing the state to pay interest to bondholders while the usina concessionaires themselves were not repaying either interest or principal. He approved Law 293 of May 9, 1898, which revoked the authority to grant loans under the laws of 1890, 1891, and 1895. Law 293 also permitted concessionaires to renew their contracts on more favorable terms of repayment: amortization would begin after the third harvest instead of the second, and at the rate of 2.5 percent for the first five years. Penalty interest would be forgiven those who promptly paid up back interest on the principal. Governor Sigismundo Antônio Gonçalves Ferreira approved Law 407 of June 28, 1899, which permitted concessionaires to rent their usinas to third parties, consolidate the debt, and forget past penalty interest; the concessionaire then had to amortize at the rate of 9 percent yearly. Governor Antônio Gonçalves Ferreira approved Law 637 of June 8, 1903, which allowed usinas in arrears to deduct a special penalty tax from their debt. During his second administration Sigismundo Gonçalves approved Law 736 of June 10, 1905, which enabled the state to sell its debts against delinquent usinas; the buyers would collect the debt as they chose, presumably by foreclosing. As a result, eleven usinas changed hands. [45]

Despite all these incentives, few concessionaires kept up their interest payments, and even fewer attempted to amortize

45. *Mensagem . . . Joaquim Corrêa d'Araujo . . . 27 de outubro de 1896*, pp. 39-40. Corrêa de Araujo also regretted that the usina loans "immobilized" capital and kept up commercial interest rates, but did not specify where capital unemployed in usina bonds might have found higher returns. The ACBP supported Corrêa de Araujo's refusal to issue more bonds. *Mensagem apresentada ao Congresso Legislativo do Estado em 6 de março de 1897 pelo Governador Dr. Joaquim Corrêa de Araujo*, pp. 50-51. *Relatório . . . ACBP . . . 9 de agosto de 1897*, pp. 13-14, 34-38, *Ibid.*, 8 de agosto de 1899, p. 79. "Lei 293," *Diário de Pernambuco*, May 11, 1898. *Mensagem . . . 6 de março de 1900 . . . Sigismundo Antônio Gonçalves. Mensagem do Exm. Sr. Desembargador Sigismundo Antônio Gonçalves Governador do Estado, lida por occasiao da installação da segunda sessão ordinaria da 5º Legislatura do Congresso Legislativo do Estado aos 6 de março de 1905*, p. 10. "Relatório . . . José de Góes Cavalcanti," *Diário de Pernambuco*, September 13, 1905. *Mensagem do Exm. Sr. Desembargador Sigismundo Antonio Gonçalves, Governador do Estado aos 6 de março de 1906*, p. 8.

the principal. By 1911, when three-quarters of the debts should have been liquidated, only four usinas had fully met their obligations. While the state Treasury had received 2,029 contos, over 90 percent had been paid in usina bonds whose real market value was 25 percent less. The outstanding debt totaled 20,426 contos, approximately one-half interest and one-half principal.[46]

The failure to collect the debts turned the loans into grants, which may have been the intention of the governors anyway. Barbosa Lima charged that Corrêa de Araujo's connections with concessionaires inhibited him from foreclosing mortgages, imposing fines, and withholding the final installment on loans as stipulated by law. The *Correio de Recife* accused Sigismundo Gonçalves of favoring political cronies with the usina transfers and of selling below actual value. A modern writer has regretted that Barbosa Lima's successors "adopted the defeatist and corrupt formula of forgiving or readjusting debts."[47]

Whatever the fiscal and political consequences of these accomodations, they at least encouraged the usinas to survive. The general success of subsidized usinas in the 1890's, in contrast with the failures of state-supported central mills in the 1880's,

46. *Mensagem do Exm. Sr. Dr.Herculano Bandeira de Mello, Governador do Estado, lida por occasiao da installação da 2° sessão ordinaria da 7° legislatura do Congresso Legislativo do Estado ao 6 de março de 1911.* Only 176 contos had been repaid in cash. Barbosa Lima's successors complained bitterly about the lack of repayment. *Mensagem . . . Joaquim Corrêa d'Araujo . . . 27 de outubro de 1896,* pp. 39-40. *Mensagem . . . 6 de março de 1900 . . . Sigismundo Antônio Gonçalves,* pp. 38-39.*Mensagem apresentada ao Congresso Legislativo do Estado em 6 de março de 1901 pelo Exmo. Sr. Dr. Antonio Gonçalves Ferreira Governador do Estado.*

47. Barbosa Lima, *Discursos Parlamentares,* v. I, pp. 383-386. In fact, Corrêa de Araujo had served as lawyer for the CSFB in the 1880's, and was a planter himself. Joaquim Corrêa de Araujo, "Publicações a Pedido, A Companhia CSFB, Limited," *Diário de Pernambuco,* February 18, 1887. Flavio Guerra, "Memórias de uma Associação (História do Comércio do Recife)," (unpublished manuscript, 1965), p. 128. See also articles critical of Corrêa de Araujo in Barbosa Lima's *A Cidade,* mid-1897, and criticisms of Barbosa Lima in Corrêa de Araujo's *Commercio de Pernambuco,* early 1899, both newspapers published in Recife. "A Administração a findar, XV-XVIII," *Correio de Recife,* February 7, 8, 10, 1908. Amaro Cavalcanti, "Antecedentes Históricos do Estatuto da Lavoura Canavieira,"*Jurídica* (Rio de Janeiro), v. xxvii, no. 70 (October-December 1962), p. 324.

can best be ascribed to four factors. First, the Republican law-makers and administrators learned from the mistakes of their Imperial predecessors and framed better laws with fewer loop-holes for fraud and non-fullfillment. Second, the system of direct loans of bonds rather than a guaranteed return proved more effec-tive, perhaps because it obliged the concessionaire to raise his capital in Brazilian securities markets and thereby to obligate him-self to local stockholders. Third, the state governments made available 15,000 contos in loans, nominally over five times the actual contribution by the imperial and provincial governments to central mills; even in real terms of the tremendously cheapened mil-réis of the later 1890's, the state loans more than doubled the value of previous subsidies. Finally, the usina functioned better than the central mill because it did not depend totally upon the cooperation of powerful independent cane suppliers. The central mill might have succeeded in an area of small and relatively poor cane growers, who had no alternative but to mill their cane in the central mill. But in Pernambuco, the senhores de engenho had made their own sugar for generations, and had thereby estab-lished an independence which they were loathe to sacrifice to the central. By retaining its own canefields the usina preserved a certain independence from the fornecedores.

The subsidized mills formed the nucleus of the modern sec-tor of Pernambuco's sugar industry. Of the 62 central mills and usinas which appeared in the state by 1910, at least 43 had received some form of subsidy. These subsidized mills alone pro-duced as much as one-third of the state's sugar crop. Moreover, in comparison with the performance of the English central mills, the Brazilian usinas obtained respectable average yields of 1,056 tons of sugar per harvest, 69.4 percent syrup-to-cane, and 7.5 percent sugar-to-cane.[48] With official aid, and despite the loss of export markets, some modernization had arrived.

This modernization left the structure of the traditional sugar

48. Fernando de Castro, "A Crise da Lavoura," *Diário de Pernambuco*, May 19, 1897. *Mensagem . . . 6 de março de 1900 . . . Sigismundo Antonio Gonçalves*, pp. 43-45. Directoria Geral de Estatistica, *Industria Assucareira, Usinas e Engenhos Centraes*, (Rio de Janeiro, 1910), p. 3.

industry intact. The new usineiros were often former senhores de engenho, and these were very often members of the ruling regional oligarchies (see Chapter Six). The organization of some usinas by joint-stock companies may have involved the entry of Recife capitalists into controlling positions on directorates, and such an influx of urban financial capital would certainly have altered the character of the sugar elite. But this influx is not evident before 1910, and remains to be studied for the period thereafter.[49]

Early signs of growing tension between traditional senhores de engenho and usineiros appear in the fights over cane supply contracts and the zoning requirements which virtually assigned plantations to specific usinas in determined sections of the zona da mata. These tensions would exacerbate later as the usinas slowly bought up engenhos and reduced their proud owners to the status of tenants and cane suppliers. But this struggle, sharpened by the temporary revival of the export trade under the impetus of the World War I and the early post-war recovery, belongs to a later period. In 1924 and again in 1929 breaks in world sugar prices moved the senhores de engenho to seek government support in the form of better cane prices and fixed production quotas.[50]

In the first thirty years of the usinas' operations, however, the tensions between the traditional and modern sectors of the industry did not aggravate to the point of forcing such accommodations. The usinas, operated in most instances by former senhores de engenho who acquired cane from other engenho owners, and by various kinds of non-propertied growers working usina lands (see Chapter Eight), led the industry in the production of sugar for both the reduced export market and the distant national market.

49. Gadiel Perucci, "Le Pernambuco (1889-1930): Contribution à l'Histoire Quantitative du Brésil," thèse de doctorat de 3ème cycle (Paris, 1972), speculates on the importance of urban financiers in usina companies.

50. Dé Carli, *O Processo Histórico da Usina*, pp. 31-40. Barbosa Lima Sobrinho, *Problemas Econômicos e Sociais da Lavoura Canavieira*, pp. 21-26.

Part Two

THE SOCIAL CRISIS

E mais: para que não pensem
que em sua vida tudo é triste,
vejo coisa que o trabalho
talvez até lhe conquiste:
que é mudar-se dêstes mangues
daqui do Capibaribe
para um mucambo melhor
nos mangues do Beberibe.

"Morte e Vida Severina," João
Cabral de Melo Neto, *Morte e
Vida Severina e Outros Poemas
em Voz Alta*, p. 112.

6

LAND:
THE BASIS OF POWER

$\mathcal{T}$he successful modernization of technology and productive relations in a group of mills did not save the industry from stagnation. Thus, capital investment alone did not solve their problems. But the planters succeeded in forcing the sugar workers to share the burden of these problems. Due to their monopolization of land, the planters could dominate the labor market. They thus derived most of the advantages of the conversion from slave to free labor and passed the costs of the economic crisis on to the workers in the form of poor compensation and poor working conditions.

The Zona da Mata

Sugar comes from sugar cane, and sugar cane grows on land. In Pernambuco, sugar planters monopolized but did not use the best lands. The total provincial area of Pernambuco comprised approximately 100,000 to 110,000 square kilometers.[1] The cane-growing area or forest zone (*zona da mata*, originally so named for the extensive forests covering the region) included the entire

1. José Bernardo Fernandes Gama, *Memórias Históricas da Província de Pernambuco*, 2 vols. (Pernambuco, 1844), Tomo I, v. I, p. 3. Jerônymo Martiniano Figueira de Mello, *Ensaio sôbre a Statística Civil e Política da Província de Pernambuco* (Recife, 1852), p. 1. These mid-nineteenth century observers estimated the provincial area between 42 and 44 leagues along the coast by 185 to 190 leagues east to west. Averaging these rough estimates, one arrives at a total area of 107,376 square kilometers. The twentieth-century state of Pernambuco includes a territory of 99,254 square kilometers. Instituto Brasileiro de Geografia e Estatística, Departamento Estadual de Estatística, *Anuário Estatística*, ano XII (Recife, 1946), p. 16.

length of the province's 170 kilometer coastline, and stretched inland 60 kilometers in the north and 130 kilometers in the south. The zone covered roughly 15,000 square kilometers, or about 16 percent of the province's area.[2]

Although the zona da mata lies on a single plain bordered in the east by the Atlantic Ocean and in the west by the Borborema Plateau, its topography, soils, and climate differ sufficiently to distinguish three major regions: tablelands, dry mata, and humid wet mata.[3] From the northern frontier south to Recife and west to Igarassú, the tablelands are cut by five perennial rivers which empty into the Atlantic. The tableland soils are sedimentary and sandy, absorb water well, and thereby insure good drainage. Along the broad river valleys the sandy soils include some clay, and the low river bank várzeas become swamps during the winter months of May through July, when 1.9 meters of rain fall at Goiana, and during high tides. Sugar cane grows well on the várzeas and half-way up the tablelands' slopes.

The dry mata extends west of these tablelands from São Lourenço da Mata and Glória do Goitá north to São Vicente Ferrer. Here the elevated plateau, or *chã*, consists of crystalline soils. Only the Goiana and Capibaribe rivers drain the area, and their flows diminish considerably during the dry season, October

2. In the late nineteenth century, Henrique Augusto Milet estimated that zona da mata covered 15,020 square kilometers. "A Colonisação," *Diário de Pernambuco*, May 17, 1888. Gileno Dé Carli estimated 14,421 square kilometers in his *Aspectos Açucareiros de Pernambuco*, p. 45. Rachael Caldas Lins and Gilberto Osório de Andrade used the figure 12,000 square kilometers, *As Grandes Divisoes da Zona da Mata Pernambucana* (Recife, 1964), p. 15, and J.M. da Rosa e Silva Neto used the figure 18,681 square kilometers in his *Contribuiçao ao Estudo da Zona da Mata em Pernambuco*, (Recife, 1966), p. 16. Our figure 15,000 is a rough average of the maximum and minimum estimates.

3. This analysis derives from Caldas Lins and Osório de Andrade, *As Grandes Divisões, passim*. Other scholars have divided the zona da mata into only two regions, the dry and the humid—Manoel Correia de Andrade, *A Terra e o Homem do Nordeste*, 2nd ed. (São Paulo, 1964, first published 1963), pp. 10-12—and the maritime and the continental, as in Vasconcelos Sobrinho, *As Regiões Naturais de Pernambuco, O Meio e a Civilização* (Rio de Janeiro, 1949), pp. 31ff. J.H. Galloway analyses two regions in terms of soils and topography in "The Sugar Industry of Pernambuco During the Nineteenth Century," *Annals of the Association of American Geographers*, v. LVIII, no. 2 (June 1968), pp. 288-290.

through December. The dry mata várzea is narrower than that nearer the coast; only 0.9 meters of rain usually fall at Pau d'Alho, and only 0.7 at Aliança. Cane grows in the várzeas, and on level areas of the chãs.

The humid mata, the principal sugar cane growing area, extends south from Recife to the Alagoas border and west to Quipapá. Drained by four large perennial rivers and their many tributaries, this area includes broad high and low várzeas near the coast and numerous narrower floodplains upriver. Although the region is quite hilly, without the large plateau of the north, it enjoys a thick fertile cover of clay soils usually called massapê.[4] Sugar cane can grow, and has grown, virtually everywhere except on the tops of the highest hills, which are left in woods for fuel and lumber. The humid mata receives up to 2.5 meters of rain along its coast, and floods are common. Even in the western interior, the rains are heavy: Catende, at a more western longitude than Aliança, receives 1.0 meter of rainfall.

Sugar production dominated the zona da mata. Between the beginning and the end of the nineteenth century, the number of sugar plantations grew from under 500 to over 2,000 (Table 21). Between the 1850's and the late 1870's, the number of mills rose from 1,300 to 1,700, a 31 percent increase which undoubtedly contributed, along with technological changes improving productivity, to the doubling of sugar production between those dates. Between the 1880's and the first decade of the present century, on the other hand, the increase in number of mills was less striking, but the quality of the new mills, some sixty of which were modern large-capacity usinas, permitted a 50 percent production increase.

In general, the number of mills increased for two reasons. First, despite the growing export market crisis, sugar remained the preferred investment in Pernambuco. Cotton briefly challenged sugar's supremacy in the period 1860-70, when the Civil

4. Gilberto Freyre, *Nordeste, Aspectos da Influência de Canna sôbre a Vida e a Paisagem do Nordeste do Brasil*, 4th ed. (Rio de Janeiro, 1967, first published 1937), pp. 6-8. Caldas Lins and Osório de Andrade object that the crystalline soils include both alluvials and colluvials—that is, soils deposited by rivers and soils eroded from hillsides—and therefore the term massapê does not indicate any one particular soil. *As Grandes Divisões*, p. 13.

TABLE 21

PERNAMBUCO SUGAR MILLS

Year	Number Identified [a]	Number Estimated
1761 [b]	230	
1775	296	
1818		500
1844	670	712
1854	532	642
1857	1,106	
1850's [c]	1,356	
1860's	1,672	
1872		1,345
1870's	1,446	
1883		2,000
1880's	1,654	
1890's	1,975	
1901		1,500
1900's	1,530	
1914 [d]	2,788	

[a] Includes plantations whose mills no longer functioned (fogo morto).

[b] We counted only those mills which remained within Pernambuco after the territorial losses of Paraíba and Ceará (1799), Alagoas (1817), and the Comarca of São Francisco (1824). Mario Melo, Síntese Cronológica de Pernambuco (Recife, 1942), passim.

[c] Figures for decades report the maximum number of engenhos identified in Pernambuco Almanacs published during those decades.

[d] Includes 500 plantations growing cane for usinas, (fornecedores).

SOURCES: For years 1761, 1775, 1843, 1850's, 1860's, 1870's, 1880's, 1890's, 1900's, 1914, see Appendix 3. 1818: Pereira da Costa "Origens Históricas," p. xxv. 1844: Fernandes Gama, Memórias Históricas da Província de Pernambuco, v. I, p. 5. 1854: Relatório . . . 1854 . . . José Bento da Cunha e Figueiredo, table. 1857: Relatório . . . 1857 . . . Sérgio Teixeira de Macedo, pp. 75-77. 1872: Report by Doyle, Parliamentary Papers, 1872, HCC, v. LVIII, AP, v. XXIII, p. 634. 1883: Report by Hughes, Pernambuco, April 8, 1883, ibid., 1883, HCC, v. LXXIII, AP, v. XXVI, p. 1,133. 1901: "A Questão do Assucar no Brasil," Diário de Pernambuco, September 26, 1901.

War in the U.S. created exceptionally favorable competitive conditions, but that challenge faded with the return of U.S. cotton to world markets. This explanation implies that profit margins in sugar, notwithstanding the high entry costs in the sugar industry, exceeded possible earnings in other sectors such as tobacco, leather, rice, or cocoa. Secondly, the technology available for cane transport and sugar manufacture did not permit building mills beyond a certain size. Once that limit was reached, the planter wishing to increase production would build a new mill. This situation only changed with the building of larger central mills and usinas with their own railroads in the final decades of the nineteenth century.[5]

Large numbers of new sugar mills appeared in the 1840's, 1850's, and 1870's, in response to particular conditions. During the 1840's new export markets opened in England, and importation of African slaves reached a peak. In 1852, the provincial president counted 137 new mills constructed in the previous decade, and in 1859 the provincial president remarked on "many new sugar mills."[6]

The extension of railroads reducing the cost of transport to Recife also encouraged engenho-building. In western areas of humid mata, such as Palmares, Gameleira, Sirinhaém, Rio Formoso, Una, Água Preta, Amarají, and Escada, through which the Recife and San Francisco Railway Company extended after 1855, the number of mills nearly doubled in the second half of the century. Similarly, the number of mills in Igarassú, Itambé, and western Goiana, where the Great Western railroad operated after 1881, grew at least 50 percent in the 1880's. Tracunhaém and Nazaré also experienced increases in engenhos, but the per-

5. The volume of Pernambuco's cotton exports rose from an annual average of 1,707 metric tons for 1856-60 to 14,405 for 1861-65,and 17,197 for 1866-70, but then declined to 13,277 for 1871-74. *Relatório . . . ACBP . . . 25 de agôsto de 1862: ibid., 8 de agôsto de 1878; Informações sôbre o Estado de Lavoura*, p. 162. Furtado, *The Economic Growth of Brazil*, pp. 121-122. Antônio Barros de Castro, *7 Ensaios sôbre a Economia Brasileira*, 2 vols. (Rio de Janeiro, 1971), v. 2, pp. 24-25 points out that although entry costs were high, the overhead costs of running a plantation were so low that once established the senhor de engenho could "go through long and deep periods of stagnation without ever encountering really critical situations." See also Canabrava, "A Grande Propriedade Rural," p. 216.

6. *Relatório . . . Víctor de Oliveira . . . 9 de março de 1852*, p. 27. Mathieson, *Great Britain and the Slave Trade*, p. 92. *Relatório com que o Excellentíssimo Senhor Conselheiro José Antônio Saraiva abrio a sessão ordinária da Assembléa Legislativa desta Província no Primeiro do Março de 1859*, p. 3.

centage change was smaller because topography and climate counterbalanced the advantages of the Great Western railroad. In areas relying on coasting vessels to carry their sugar to Recife, such as Ipojuca, Barreiros, and eastern Goiana, on the other hand, the number of mills did not grow noticeably. Nor did engenhos increase in regions of limited land or encroaching urban populations, such as the island of Itamaracá, the Recife suburbs of Afogados, Várzea and Poço da Panela, and nearby Olinda (Appendix 3).[7]

While the sugar planters owned most of the zona da mata, they used productively very little of it. In the 1850's, the planters probably used no more than one-fifth of available lands, if the following two calculations are accurate:

1. Given industrial yields of five kilograms of sugar from every 100 kilograms of cane ground (hereafter expressed as 5 percent sugar-to-cane), the Pernambuco mill owners would have had to grind 1,280,500 tons of cane to produce the 1854 harvest of 64,025 tons of sugar. Given agricultural yields of 40 tons of cane per hectare planted, the plantation owners would have had to harvest 32,013 hectares to yield that much sugar cane. Sugar cane matured in twelve to seventeen months, so to insure an annual crop on 32,013 hectares, the senhores de engenho had to plant twice that area. They also had to assign land to pasturage for work animals and food crops, a total perhaps equal to that planted and reserved for sugar cane.[8] Thus, the amount of land

7. Recife and San Francisco Pernambuco Railway Company (Ltd.), *Report of the Proceedings at the Thirty-First Half-Yearly Ordinary General Meeting of the Shareholders* . . . *December 13, 1871*, p. 3. *Idem., Report* . . . *Forty-Eighth Half-Yearly Ordinary General Meeting* . . . *October 7, 1879*, p.2.

8. *Relatório apresentado à Assembléa Legislativa Provincial em o 1º de março de 1864 pelo Exm. Sr. Dr. Domingos de Souza Leão vice-presidente de Pernambuco*, p. 24. This source estimated sugar-to-cane yields at 4, 5, and 6 percent. Peres and Peres, *A Industria Assucareira em Pernambuco*, p. 83, quoted a nineteenth-century French engineer in Pernambuco who gave 4 percent as an upper limit. For production data, see Table 5. The earliest available estimates of agricultural yields in Pernambuco at the end of the nineteenth and beginning of the twentieth century mention 60 tons per hectare most frequently. Raffard, *O Centro da Industria*, p. 112. Walle, *Au Brésil du Rio São Francisco à l'Amazone*, p. 164. Manoel Antônio dos Santos Dias Filho, "Industria Assucareira," *Boletim do Ministério da Agricultura, Industria e Commercio* (Rio de Jan-

utilized for sugar production in 1857 would equal 96,039 hectares.

2. The relatively prosperous sugar parishes of Ipojuca and Jaboatão produced an average of 63.7 tons of sugar per engenho in 1854. Using the same assumptions as in estimate 1, one finds that each mill would have to harvest 31.9 hectares yielding 1,274 tons of cane to manufacture this sugar. Then the 1,269 engenhos counted in the province in the 1850's would have had to harvest 40,481 hectares, and use a grand total of 121,443 hectares.[9]

Modern Pernambuco geographers estimate that 37.5 percent of the zona da mata, or 574,700 hectares, can be planted in cane. If one assumes pasturage and food crops were also restricted to this portion, then we can conclude that mid-nineteenth century planters used between 16.7 percent (estimate 1) and 21.1 percent of the zona da mata's cultivable land (estimate 2).[10]

Various motives explain why sugar planters used so little of the available land. Limitations in transportation facilities and mills restricted the amount of cane any one mill could transform into sugar, and thereby restricted plantings. After a certain distance from the mill, cane cost too much to carry by animal. The cane had to be milled within 48 hours after cutting to extract usa-

---

eiro), Anno I, no. 5 (1913), p. 60. The Directoria Geral de Estatistica, *Indústria Assucareira, Usinas e Engenhos Centraes,* p. 3, calculated a return of 40 tons per hectare. We accept the latter figure since it tends to increase our estimate of the amount of land under cultivation and thus works against our hypothesis, which stands nevertheless. Using 60 tons per hectare would render the proportion of utilized land still smaller. Tollenare, *Notas Dominicais,* p. 72. Compare Dé Carli, *O Processo Histórico da Usina,* pp. 60-61, who figured 74 percent of cane lands were necessary for pasturage, and an equal amount for wood reserves.

9. *Relatório . . . 1854 . . . José Bento da Cunha e Figueiredo,* Table.

10. Caldas Lins and Osório de Andrade, *As Grandes Divisões,* p. 15. If we assume that land unusable for cane could serve for pasturage or food crops, the percentage of cane lands utilized drops even lower. Compare two early nineteenth-century observers: Tollenare, who estimated that a planter used only 8.6 percent of his land, and Henry Koster, who remarked that "an estate contains in general much more land than its owner can manage or in any way employ." *Notas Dominicais,* pp. 71-72. *Travels in Brazil,* v. II, p. 135. See also Dé Carli, who found that 13.1 percent of the total zona da mata was planted in cane for the 1929-30 harvest. *O Processo Histórico da Usina,* p. 65.

ble syrup, which meant in the pre-usina era that even if the cost of animal transport were acceptable, cane cut and stored for more than two days was not worth milling. The engenhos themselves could usually grind no more than 25 tons of cane daily, and the harvest begun in September ended with the early rains in January or February which lowered cane saccharose content. Only significant rises in world prices would justify a longer milling season and grinding inferior-grade cane.[11]

Planters also kept land idle for its potential value. Given the fluctuating and usually inflating national currency, the wise planter invested in real estate. He also used his idle land to maintain a free population of sharecroppers, tenants, and squatters, who grew food crops for their own or the engenho's consumption, or grew sugar cane. This subservient free population served a multitude of economic and political purposes; their dependency upon the senhor de engenho derived from his ownership of the land and his power to evict them at a moment's notice.[12]

A few individuals not only recognized the disadvantages of these uncultivated lands but also suggested reforms. They called for real estate taxes scaled to weigh heavily on uncultivated lands, and most heavily on uncultivated lands near public roads and railroads, where the state had spent money to encourage a private initiative which did not respond. These reformers argued that real estate taxes, unlike export taxes and excise taxes, did not penalize the producer for producing more, nor the consumer for consuming more. But the sugar planters, most of whose lands

11. Galloway, "The Sugar Industry of Pernambuco," p. 291. Souza Leão Pinto, *Cana-de-açucar*, p. 22. The 532 mills described in 1854 produced an average of 178 tons of sugar per harvest, or around 1.3 tons per day. This quantity of sugar required grinding 22.6 tons of cane each day, assuming 5 percent sugar-to-cane. *Relatório . . . 1854 . . . José Bento da Cunha e Figueiredo*, Table. The 1857 report on 1,106 mills gave an average of only 71 tons per harvest, or one-half ton of sugar from 10 tons of cane per day. *Relatório . . . 1857 . . . Sérgio Teixeira de Macedo*, pp. 75-77. Koster, *Travels in Brazil*, v. II, p. 116, and Tollenare, *Notas Dominicais*, p. 78, give harvest months.

12. See Andre Gunder Frank, *Capitalism and Underdevelopment in Latin America: Historical Studies of Chile and Brazil* (New York, 1967), pp. 258-259. Celso Furtado, *Economic Development of Latin America. A Survey from Colonial Times to the Cuban Revolution*, translated by Suzette Machado (Cambridge, Eng., 1970), pp. 226-227.

were uncultivated, had little sympathy for such proposals, and the appeals failed.[13]

Escada

To examine more closely the nature and consequences of the planters' land tenure, one can study an individual sugar area. A review of data on population, sugar production, and number and size of rural properties in the early twentieth century revealed that Ipojuca, Jaboatão, Cabo, Igarassú, Rio Formoso, Goiana, São Lourenço da Mata, Sirinhaém, and Escada shared similar characteristics.[14] We chose Escada because of the quantity and quality of data available. (See Plate 6.)

Originally an Indian village under the protection of the Portuguese authorities in the later seventeenth century, Escada was first settled by Brazilians of European descent in the early eighteenth century. The area became a parish in 1786 and gained its own local government and the status of village in 1854. In 1868 the Indians were permanently removed, and in 1873 the village received its own judges and the status of town. In 1893, Escada became an independent county subordinate directly to the state government.[15]

Fifty-eight kilometers southwest of Recife, Escada covered some 625 square kilometers in the humid mata "where in large part the lands are hilly and mountainous . . . the flat lands are small and there are few várzeas." The Ipojuca River ("man-kill-

13. "O dr. José Antonio de Figueiredo ao publico, XVI, XVII," *A Província*, January 28, 30, 1875; "Collaboração Breves Considerações sobre a agricultura no Brasil," *Diário de Pernambuco*, May 2, 1876; SAAP, *Trabalhos do Congresso Agrícola*, pp. 217, 221.

14. These characteristics were mean populations in 1910, 1920, and 1930 between 10,000 and 30,000; similar numbers of proprietors, renters, and properties classified into six categories by size in 1920; and three or more larger usinas by 1929. *Recenseamento do Brazil realizado em 1 de Setembro de 1920*, v. III, part 1, "Agricultura." Directoria Geral de Estatística, *Relação dos proprietários dos Estabelecimentos Ruraes Recenseados no Estado de Pernambuco* (Rio de Janeiro, 1925). Gonçalves e Silva, *O Assucar e o Algodão. Annuário Estatístico de Pernambuco*, Ano IV (1930), all *passim*.

15. Samuel Carneiro Rodrigues Campello, *Escada e Jaboatão: Memória apresentada ao VI Congresso de Geographia Brasileiro*, (Recife, 1919), pp. 3-5. F. A. Pereira da Costa, "Origens historicas do Município da Escada," *A Escada*, April 17, 1904.

er" in the Indian language, so called for its treacherous and tur-
bulent waters) with sources in the sertão, traversed the parish
from west to east to flow into the Atlantic Ocean 40 kilometers
south of Recife. In 1852, a provincial judge wrote: "The waters of
the Ipojuca are clear and healthy. Until Escada [traveling down
stream] it can hardly be forded, and cannot be navigated due to
many rocks and a few water-falls. . . . From one port to another
some small rafts still travel on this river."[16]

A few large sugar plantations owned by a small group of families dominated Escada. Among 84 sugar plantations surveyed in
Escada in the 1850's, only 15 percent held more than 3,000 hectares, but these included 70 percent of all surveyed plantation
land. Conversely, plantations under 3,000 hectares constituted
85 percent of the sugar properties but held only 30 percent of the
plantation lands. Thus the median plantation size—the size larger than one-half the group and smaller than the other half of the
group—was only 995 hectares, although the average plantation
was 2,871 hectares. Much smaller were 55 other farms (sítios,
propriedades) which produced food crops and measured only 350
hectares each. Moreover, if one excludes those properties which
became sugar plantations before 1880, the average non-sugar
property size drops to 166 hectares, or about one-twentieth the
size of the average plantation.

Data from two other sugar parishes in 1857 confirm this concentration of land in a few hands. In Sirinhaém southeast of
Escada, four families produced 37 percent of the sugar from a
group of 73 engenhos, and in Nazaré in the northern dry mata six
families controlled 57.4 percent of total sugar production.[17]

In Escada, a small clique of eight interrelated families ruled

16. The total measured areas of Escada and Amarají, which became a
separate county in 1889, were 623 square kilometers. Instituto Brasileiro de
Geografia e Estatistica, Departamento Estadual de Estatistica, Anuário
Estatistico, ano XI (1942), Melo, Síntese Cronológica, p. 111. Quotes in
Campello, Escada e Jaboatão, pp. 5-6, and Figueira de Mello, Ensaio
sôbre a Statística, p. 9.
17. Vasconcellos, Almanack . . . 1854, 1860, 1861. APE, Registro de
Terras Públicas, v. V (1858-59), v. VII (1858-78). Ibid., Registro de
legitimação de posse de terras (1872-75), contains land surveys. Ibid., Coleção
Camaras Municipaes, 1857.

as the landed oligarchy. The most prominent Escada family, the
Lins, was descended paternally from a Portuguese who immigrat-
ed to Brazil in the earlier eighteenth century. This man's son
obtained land in Sirinhaém in 1813, and ten years later Henrique Marques Lins, a grandson, received a sesmaria, an official
but by then illegal grant of one square league adjoining his
father's plantation. In 1841, with the income from that estate,
Henrique bought the Engenho Conceição in Escada, and also the
land upon which he built the Engenho Matapiruma, which first
ground cane in 1842. During the last forty years of the century,
Henrique Marques Lins, who received the titles of Barão and
later Visconde de Utinga, and his nine children and their spouses
owned at least thirty different mills located in Escada. In 1877
and 1881, Henrique Marques de Holanda Cavalcanti, a grandson
and later Barão de Suassuna, built the modernized Usinas Mameluco and Limoeirinho, and in 1895 Marcionilo da Silveira Lins,
a son, inaugurated the Usina Massauassú.[18]

Next in number of plantations came the Pontual family, led
by José Manoel Antônio Pontual, his son José Manoel Júnior, and
various cousins and in-laws. The Pontuals owned seventeen
engenhos and one sítio; they built the Usinas Aripibú and Cabeça de Negro in 1888, and the Usinas Bosque and Mussú in the
1890's. The Santos family, led by the patriarchs José Rodrigues
de Senna Santos and João Felix dos Santos, owned sixteen engenhos in the later Empire, while the Velloso da Silveira clan of José
Pedro Velloso da Silveira and his sons Eustáquio and Cincinato
held twelve engenhos and one sítio in Escada. José Pereira de
Araújo and André Dias de Araújo, later Barão de Jundiá, owned
eleven plantations, and José Pereira built the Usina Bamburral in
1888. Manoel Antônio Dias and his sons José Cândido and Manoel
Antônio dos Santos Dias owned nine engenhos; in 1889 Santo Dias
built the Usina Santa Philonila and in 1895 he put the Usina Bom
Fim into operation. José Antônio Barros e Silva and his brother

18. Pernambuco, Biblioteca Pública, *Documentação Histórica Pernambucana, Sesmarias,* 3 vols. (Recife, 1959), v. IV. p. 36. Volume III has not yet
been published. The sesmarias were formally suspended in 1822. Costa Porto,
Estudo sôbre o Sistema Semarial, p. 172. Guilherme Auler, *Os Utinga. Filhos,
Netos e Bisnetos do Senhor do Engenho Matapiruma* (Recife, 1963), pp. 12-19.

Francisco Antônio, later Barão de Pirangy, also owned nine mills. Finally, both the Alves da Silva and Siqueira Cavalcante families owned five engenhos each.[19]

The sum of 114 engenhos among the nine families reflects every mill owned during the period, not necessarily the total held at any one time. The sum thus includes repetitions of property transferred within the group and does not exclude property transferred to non-oligarchy planters. Moreover, many families held property outside Escada which extended their influence and may have changed their ranking by total number of properties. Despite these qualifications, however, in a county of some 120 mills the oligarchy clearly held the great majority, and it built all the modern usinas.[20]

The Escada sugar oligarchy protected and expanded its power by controlling local politics. Sugar planters dominated both administrative and judicial branches of local government and enjoyed representation in the Pernambuco legislature. In 1861

19. We have avoided the temptation to try to establish kinship through analyzing surnames. Normally the child received the last maternal surname followed by the last paternal surname. Thus Davino dos Santos Pontual's mother was probably a Santos and his father was probably a Pontual. Unfortunately for amateur genealogists, however, this procedure was frequently ignored, and the limited variety of surnames allowed individuals with the same name to be unrelated to each other. Thus, for example, although five individuals with the paternal surname Ferreira owned a total of eleven engenhos, none of them shared a maternal surname and we cannot assume kinship. Upon marriage, moreover, a woman chose which surnames she would add and retain according to aesthetic, sentimental, or practical criteria. Our kinship references, when not obvious, are from Auler, who consulted family records. More remote Lins ancestors came to Brazil in the sixteenth and seventeenth centuries. Henrique Oscar Wiederspahn, "Dos Lins de Ulm e Augsburgo aos Lins de Pernambuco," *Revista do Instituto Arqueológico, Histórico e Geográfico Pernambucano*, v. XLVI (1961), pp. 7-98; see also Carlos Xavier Paes Barreto, "A estirpe dos Lins," *ibid.*, pp. 209-215.

20. Lists of engenhos and their owners can be found in Vasconcelos' *Almanack* . . . *1860, 1861, 1862;* and in Amaral's *Almanack* . . . *1869, 1870, 1872, 1873, 1875, 1876, 1881, 1884, 1885, 1886;* in Veríssimo de Toledo's *Almanack* . . . *1893, 1894, 1895;* in Pires Ferreira's *Almanack* . . . *1900, 1902;* and in *Annuario Commercial Pernambuco, Parahyba, Alagôas, Bahia, 1902-03,* (Recife, 1903). Obvious omissions in these listings prevented specifying property transfers, which may have numbered around 30 engenhos acquired by the oligarchy and another 30 ceded to non-oligarchy planters.

the Escada seven-man town council included Henrique Marques Lins and his son Belmiro da Silveira Lins, later Barão de Escada, and José Pereira de Araújo, men who owned a total of thirteen plantations. In 1881, a nine-man town council had three planters with seven mills, and in 1893 the mayor owned five mills and five councilmen or their children owned fourteen mills.

The police chief and his deputies, assisted by judges, maintained law and order and administered justice. In Escada the police chief in the early 1860's was Henrique Marques Lins' son-in-law, Antônio Marques de Holanda Cavalcanti, aided by district deputies Francisco Antônio de Barros e Silva and João da Rocha Holanda Cavalcanti, the former a brother-in-law to Henrique Marques Lins, and the latter probably the police chief's brother. These three police officials owned a total of nine engenhos. In the 1880's the police chief, Samuel dos Santos Pontual, and three assistants owned a total of thirteen sugar mills.

Escada municipal and district judges during these years were not local sugar planters, although they occasionally came from sugar families of neighboring counties. In the 1880's two of the municipal judge's three substitutes were an Araújo and a Santos Pontual, with eleven sugar mills between them. The justices of the peace, on the other hand, were almost exclusively senhores de engenho. The seventeen justices of the peace in the early 1860's included twelve senhores de engenho with a total of nineteen sugar plantations; in the 1880's, twelve of the sixteen justices of the peace were mill owners or mill owners' sons, with a total of eighteen mills.

Two National Guard battalions stationed in Escada constituted an additional force for social control; the national guard was also the source of the titles "colonel" "major" and "captain," so proudly used by sugar plantation owners. In the 1860's, Lieutenant Colonel Henrique Marques Lins commanded the 24th Battalion, and Lieutenant Colonel Manoel Gonçalves Pereira Lima, the owner of two engenhos, commanded the 25th Battalion. The fifteen company captains in these battalions included eight sugar plantation owners with a total of sixteen sugar mills. In the 1870's, all sixteen of the company captains were sugar mill owners, and in the 1880's Colonel André Dias de Araújo commanded the National Guard, and Lieutenant Colonel Antônio dos Santos Pontual and Lieutenant Colonel José Pereira de Araújo led the battalions.[21]

At the provincial and state level in government, the Escada sugar oligarchs shared power with representatives of similar elites. Several Escada planters served two-year terms in the 36-member provincial assembly: João Francisco de Arruda Falcão and José Pedro Velloso da Silveira in 1851; the former and Samuel dos Santos Pontual in 1861 and 1867 respectively; the latter in 1877; Francisco Dias de Arruda Falcão in 1881 and 1883, André Dias de Araújo in 1885, and Davino dos Santos Pontual in 1887. In the new Republic, Davino dos Santos Pontual participated in the state constituent assembly, Luis Caldas Lins was a member of the first state legislature of 1892-94, and João Alves Pontual was elected to the state legislature from 1904 to 1912. Sugar planters from other counties also appeared frequently in lists of provincial and state assembly representatives.[22]

Challenges to Planter Hegemony

Within Pernambuco, therefore, the sugar planters' political supremacy was nearly absolute. The most serious threats to this dominance came not from other classes subject to their power, but rather from within the planter class itself. The senhores de engenho belonged to both Liberal and Conservative political parties during the Empire, and political differences over such questions as elections and electoral requirements, senators' life tenure, the imperial power to dismiss cabinets and legislatures

21. Lists of officials can be found in the *Almanacks* edited by Vasconcelos, Amaral, and Verissimo De Toledo, 1860-95.

22. Netto Campello, *História Parlamentar de Pernambuco* (Recife, 1923), pp. 111-151. Beyond Pernambuco, sugar planters appeared less frequently in lists of officials, and their behavior was less obviously determined by narrow self-interest. Nevertheless Escada planters sent politicians to the Rio de Janeiro government. For a list of such men and their posts and tenure, see Peter L. Eisenberg, "The Sugar Industry of Pernambuco, 1850-1889," Ph.D. dissertation at Columbia University (New York, 1969), pp. 94-95; and Dunshee de Abranches, *Governos e Congressos da República, 1889-1917*, 2 vols. (Rio de Janeiro, 1918), *passim*.

(*poder moderador*), government control of the press, and the distribution of power between imperial and provincial governments, as well as personal questions, sometimes led to bloody disputes.[23]

The Beach Revolution (Revolução Praieira), the last major political convulsion in Pernambuco in the Second Empire, illustrates the point. This movement began when the Liberals resisted the return of the Conservative Rêgo Barros and Cavalcanti families to provincial power. Led by the Barão de Boa Vista (Rêgo Barros) and the Viscondes de Camaragibe, Suassuna, and Albuquerque (three Cavalcanti brothers), these families had dominated Pernambuco politics during the years 1837-44, and they expected to return when the Emperor restored the Conservatives to power in September 1848. This expectation was disappointed when the Liberal Praieiros, joined by "the inferior and lower and ignorant classes of the population who consider themselves disinherited of social wealth or oppressed by tyrannical and offensive laws" revolted against the Rêgo Barros, the Cavalcantis, and the foreign merchants who dominated commerce in the cities.[24]

A rich literature of pamphlets, magazines, and newspapers publicized the Liberals' grievances. Antônio Pedro de Figueiredo, a mulatto and one of the most forceful of these writers, castigated the planter oligarchy. "These modern feudal barons, when their estates are located far from the provincial capital, live in an almost complete independence, administering justice with their own hands, arming their vassals for open warfare with one anoth-

23. For surveys of differences between Liberals and Conservatives, a topic still awaiting serious treatment, see João Camillo de Oliveira Torres, *A democracia coroada: teoria política do império do Brasil* (Rio de Janeiro, 1957); *idem.*, *Os Construtores do Império, Ideais e lutas do Partido Conservador Brasileiro* (São Paulo, 1968); and Iglesias, "Vida Política, 1848-1888," pp. 9-132.

24. The revolt was named after Recife's Beach Street, where the rebels published the principal newspaper. Flávio Guerra, *História de Pernambuco*, 2 vols. (Rio de Janeiro, 1966), v. II (*Provincia e Estado*), p. 35. Amaro Quintas, "O Nordeste 1825-1850," in Buarque de Holanda, *História Geral da Civilização Brasileira*, Tomo II, v. 2, *Dispersão e Unidade*, pp. 226-227. Jerônymo Martiniano Figueira de Mello, *Chrônica da Rebelião Praieira* (Rio de Janeiro, 1850), p. 4, quoted in Amaro Quintas, *O Sentido Social da Revolução Praieira* (Rio de Janeiro, 1967), p. 39.

er, in disobedience of the orders of the government and the decrees of the judges." Figueiredo called for the end of "the large landed estate," and the establishment of a rural middle class, through direct taxation and real estate taxes which "would force the large landowners to get rid of the lands which are of no use to them."[25]

Despite Antônio Pedro de Figueiredo and the popular feelings against the Rêgo Barros and Cavalcantis, the rebels did not attack all members of the sugar oligarchy. Numerous senhores de engenho joined the Praieiros. From Escada, Francisco José de Barros e Silva, and João Félix dos Santos, both powerful planters owning many engenhos, joined the revolt, and Escada engenhos such as Freixeiras and Camassari were fortified and used as supply depots for the revolutionary forces. Pedro Ivo Velloso da Silveira, presumably a relative of Escada oligarch José Pedro Velloso da Silveira, led the revolutionary army which almost captured Recife.[26]

The Praieiros may have been "the incarnation of popular soul, sentiment, and aspiration," but the Revolução Praieira, contrary to recent interpretations, threatened only that part of the sugar oligarchy which, through its domination of provincial politics one decade earlier and its intention to resume that domination in 1848, had made itself politically most odious.[27] The revolt itself lasted three months, from November 1848 to January 1849. The provincial authorities, aided by imperial forces, defeated the rebel armies in two decisive battles and crushed the movement.

25. "Pernambuco, Revista retrospectiva," *O Progresso*, Tomo I, p. 298, quoted and translated in T. Lynn Smith (editor), *Agrarian Reform in Latin America* (New York, 1965), p. 74. "Colonisação do Brasil," *O Progresso*, Tomo II, pp. 634, 637, quoted and translated in Smith, *Agrarian Reform in Latin America*, pp. 69, 71.

26. Oliveira Lima, *Memorias (Estas minhas reminiscências. . .)* (Rio de Janeiro, 1937), pp. 120-121, supports the kinship presumption. Barbosa Lima Sobrinho (Alexandre José Barbosa Lima Sobrinho), *A Revolução Praieira* (Recife, 1949), p. 30. Edison Carneiro lists at least ten senhores de engenho among the principal participants, in his *A Insurreição Praieira (1848-1849)* (Rio de Janeiro, 1960), pp. 175-179.

27. Joaquim Nabuco in *A Província*, February 2, 1898, quoted in Quintas, *O Sentido Social*, p. 125. Quintas overemphasizes, in our opinion, the social revolutionary aspects of the movement.

The bitterness of the Praieira broke out again in 1880, when Liberals and Conservatives contested the succession to a senator's seat left vacant by the death of Francisco de Paula Cavalcanti de Albuquerque, Visconde de Suassuna. On election day in Vitória de Santo Antão, the deceased's district, Liberals led by the Souza Leão family, powerful planters in Jaboatão, occupied the polls with police and gunmen. Another armed force of Conservatives and dissident Liberals rode into town, and fighting erupted. Sixteen men died, including Escada planters Belmino da Silveira Lins and José Pedro de Oliveira; the wounded included the dead Belmino's father, Henrique Marques Lins, and his brother-in-law Ambrósio Machado da Cunha Cavalcanti. The battle became known as the Vitória Hecatomb.[28]

Besides political questions, antagonisms within the sugar oligarchy developed from disputes over land titles, wills, women, livestock, and slaves. While often bitter and sometimes involving violence, these disputes concerned only personal differences between individuals and families. The antagonisms did not pit one social class against another nor one economic interest group against another. Recent studies have emphasized that the actual agents of much of this violence, hired gunmen known as *jagunços, capangas,* and especially *cangaceiros,* were themselves more or less conscious social protesters. This literature emphasizes that the cangaceiro was an oppressed individual who became a professional outlaw as the best means of coping with a hostile social environment.[29] In Pernambuco, however, the can-

28. Not a cholera epidemic, as mistakenly noted by Samuel Putnam in his translation of Gilberto Freyre's *The Masters and the Slaves,* p. xlviv. "A hecatombe de Victoria," *Diário de Pernambuco,* June 30, 1880. *Falla . . .Franklin Américo de Menezes Dória . . . 1 de março de 1881,* pp. 4-8. Melo, *Síntese Cronológica,* pp. 104-105. Diogo de Melo Meneses and Gilberto Freyre (editors), *O Velho Félix e suas "Memórias de um Cavalcanti"* (Rio de Janeiro, 1959), pp. 58-65. The Vitória Hecatomb violence was not typical: another Pernambuco senhor de engenho characterized provincial politics under the Empire as peaceful. "The political disputes did not separate families under the monarchy. The party preferences did not interfere with personal relations," Júlio Bello, *Memórias de um Senhor de Engenho* (Rio de Janeiro, 1938), p. 194.

29. Rui Facó, *Cangaceiros e Fanaticos,* 2nd ed. (Rio de Janeiro, 1965); Amaury de Souza, "The 'Cangaço' and the Politics of Violence in Northeast

gaceiro usually kept to the agreste and sertão regions, and only occasionally invaded the western areas of the zona da mata. The most famous Pernambucan cangaceiro before 1914, Antônio Silvino, once attacked Manoel Antonio dos Santos Dias' Usina Santa Philonila in an attempt to carry off the estranged wife of the usineiro's son-in-law. The attempt failed, but the cangaceiros accidentally killed the usineiro's youngest daughter, a crime Silvino publicly regretted when finally captured. The *cangaço*, or outlaw life style, did not, in the other words, constitute a menace to the sugar oligarchy; on the other hand, both contemporary newspapers and the popular poetry *(literatura de cordel* or *folhetos)* report that the cangaceiro often shared his spoils with or offered protection to the poor.[30]

Two other groups might have opposed the planter oligarchy: the miniscule middle class and the urban merchants. The first group however, rarely articulated a political interest. A few isolated journalists and lawyers, in the spirit of Antônio Pedro de Figueiredo, raised criticisms in Escada. In 1863, the editor of the first and only edition of *O Escadense* attacked the Lins family:

Escada was considered the fief of Mr. Utinga, and the noble families and distinguished citizens who live in this historic locality were considered

Brazil," in Ronald H. Chilcote (editor), *Protest and Resistance in Angola and Brazil. Comparative Studies* (Berkeley and Los Angeles, 1972), pp. 110-131. Ralph della Cava, *Miracle at Joazeiro* (New York, 1970) E.J. Hobsbawm, *Bandits* (Harmondsworth, Eng., 1972; first published 1969).

30. "Usina Santa Filonila," *Diário de Pernambuco*, October 11-18, 1899. "Odysséa de um Bandido," *Jornal do Recife*, December 2, 1914. "O cangaceiro, Balisa," *Diario de Pernambuco*, August 23, 1906. "Antonio Silvino. No Pilar, notas minuciosas," *ibid.*, March 5, 1907. "Guerra aos bandidos.," *Pernambuco* (Recife), February 26, 1909. Francisco das Chagas Baptista, "A Política de Antonio Silvino;" Francisco Alves Martins, "Antonio Silvino e o Negro Currupião;" Luis de Lira, "As Bravuras de Antonio Silvino em honra de um velho amigo;" and Manoel Camilo, "O Grande e Verdadeiro Romance de Antonio Silvino," are some of the folhetos extolling Silvino's virtues. Mario Souto Maior, *Antônio Silvino, Capitão de Trabuco* (Rio de Janeiro, 1971), pp. 34, 46, recognizes that Silvino was a product of his environment and had a sense of moderation; Souto Maior also notes that Luis da Câmara Cascudo, *Flor dos Romances Trágicos* (Rio de Janeiro, 1966) and characters in José Lins do Rego's *Fôgo Morto* (Rio de Janeiro, 1943), shared this opinion. For a different view, based on folhetos, see Leandro Mota, *Violeiros do Norte*, 3rd ed. (Fortaleza, 1962), pp. 240-252.

as so many slaves of Mr. Utinga's family. It makes one blush to see an old illiterate and a family of illiterates, led and advised by an old fox, collecting among themselves all the power in the parish, and imposing themselves upon the parish by the most infamous and shameful means of violence, fraud, and corruption.

The writer credited Utinga's power in part to his godson, former (1853-56) provincial president José Bento da Cunha e Figueiredo, for "the posts of command in the National Guard, the police delegate and his substitutes, the subdelegates of the various districts and their substitutes, the municipal judge's substitutes, the justices of the peace, the president and councilmen in the town council, all came to be monopolized by the family of the happy godfather of that president."[31] In an even sharper tone, Escada's most famous resident intellectual and lawyer, Tobias Barreto de Menezes, escoriated "a *sugarocracy* . . . a *handpicked* nobility, for the most part stupid and pretentious, and even worse than the clergy, for the latter at least do not whip citizens or put them in stocks in the engenhos," and lamented that the political center of gravity in the plantations left the town of Escada "a branch of the estates" and its residents "travelers who came together at night in the same inn."[32] This middle class participated actively in the abolitionist campaigns of the 1870's and 1880's, where it reinforced an initiative announced by the Liberal Party in its 1869 program. But·the abolitionist campaigns did

31. "Situação política da Escada," *O Escadense; Periódico Político*, July 17, 1863. Campello, *Escada e Jaboatão*, p. 11.
32. Tobias Barreto de Menezes, "Um Discurso em Mangas de Camisa," in Hermes Lima, *Tobias Barreto (A Época e o Homem)* (São Paulo, 1939), pp. 289, 294, 296-297. Tobias had personal reasons as well for hating the "sugarocracy": he had fallen in love with the daughter of a Cavalcanti who prohibited the marriage because Tobias was colored. He subsequently married a daughter of João Felix dos Santos, the Escada planter, and lived there for ten years in the expectation of a dowry and later an inheritance, neither of which he received. *Ibid.*, pp. 13, 22-24, 30-31. For a less critical view of the oligarchy, see Gilberto Freyre, *Vida Social no Brasil nos Meados do Século XIX*, translated by Waldemar Valente (Recife, 1964). This study, originally Freyre's Master's Thesis at Columbia University, was first published in English as "Social Life in Brazil in the Middle of the Nineteenth Century," *Hispanic American Historical Review*, v. V (1922), pp. 597-630.

not threaten planter hegemony because by that time, as will be shown in following chapters, the planters had already substituted free labor.[33]

The Recife merchants chose not to challenge the planters politically. "A people does not live on politics; to be liberal or conservative does not matter to the economic and social regime of the country, whose progress depends more on commercial and agricultural prosperity in the counties, than on the *meditations* of the directing clubs, and the speeches of the demagogues." This aloof business-oriented mentality probably developed from an identity of interests: the major merchants themselves were sugar exporters and importers whose best clients were senhores de engenho. Many of the merchants, moreover, were foreigners reluctant to antagonize the host country: during the Praieira, for example, 46 of Recife's 77 biggest merchants were aliens, mostly English and Portuguese.[34]

The major political change in the later nineteenth century, the fall of the Empire and the advent of the Republic in 1889, did not shake the planters' supremacy. They welcomed the Republic, in part from bitterness for the imperial government's having freed their slaves without indemnization, but also possibly for fear that abolitionist influence in the Empire would promote land reform. The tiny middle class participated in the republican movement, but the planters retained control of local politics and were able to persuade the early state governments to subsidize modernization.[35]

33. Coriolano de Medeiros, "O Movimento da Abolição no Nordeste," in *Livro do Nordeste Commemorativo do Primeiro Centenário do Diário de Pernambauco 1825-1925* (Recife, 1925), p. 93. Haring, *Empire in Brazil*, pp. 96-97.

34. Associação Commercial Beneficente, "O commercio e as eleições," *Diário de Pernambuco*, February 26, 1881. Quintas, *O Sentido Social*, p. 24. On the English in Recife commerce, see Estevão Pinto, *A Associação Comercial de Pernambuco. Livro Comemorativo de seu primeiro centenário (1839-1939)* (Recife, 1940), pp. 12-14; Alfredo J. Watts, "A Colônia Inglesa em Pernambuco," *Revista do Instituto Arqueológico, Histórico e Geográfico Pernambucano*, v. XXXIX (1944), pp. 163-170; Gilberto Freyre, *Ingleses no Brasil: aspectos da influência britânica sôbre a vida, a paisagem e a cultura do Brasil* (Rio de Janeiro, 1948); and Graham, *Britain and the Onset of Modernization*, p. 74.

35. Richard Graham, "Landowners and the Overthrow of the Empire,"

While the sugar interests dominated the economy, many felt a growing need to form class associations to defend their interests after the first quarter of the nineteenth century. This need developed in part from purely practical considerations, such as the utility of organizing sugar trading and keeping statistics, but it also came from the increasing awareness, provoked by the export crisis, that the industry required technological improvements to remain competitive. The Agricultural Commercial Association (ACA), founded in Recife in 1836, was the province's oldest merchants' organization. This group included sugar warehousers, exporters, and correspondentes; its major function was to act as the sugar exchange, where prices were listed and sales conducted. Portuguese merchants dominated the ACA directorates. The Pernambuco Beneficent Commercial Association (ACBP), founded in Recife in 1839, represented commerce in general. Its annual reports pleaded for abolition of interprovincial taxes, reductions in import and export duties, and compensated abolition of slavery, and they included reviews of sugar and cotton markets. Portuguese merchants also dominated the ACBP, with a few English traders serving occasionally as presidents.[36]

The Imperial Institute of Agriculture of Pernambuco (Imperial Instituto Pernambucano de Agricultura, IIPA), the first provincial organization directly concerned with agriculture, was founded in 1860, one year after the Emperor Pedro II visited Pernambuco.[37] Despite imperial and provincial subsidies for educational ativities and experiments, the IIPA languished: the provin-

---

Luso-Brazilian Review, (Madison, Wisconsin) v. VII, no. 2 (December 1970), pp. 46-47. A study of the republican movement in Pernambuco remains to be written; the role of the middle class has been suggested by Nicia Vilela Luz, "O papel das classes médias brasileiras no movimento republicano," *Revista de História* (São Paulo) anno XV, v. XXVIII (1964), pp. 13-27.

36. Pinto, *A Associação Comercial de Pernambuco*, pp. 212-218.

37. Vasconcellos, *Almanack. . . 1862*, p. 27. *Relatório com que o Exm. Sr. Comendador Dr. Domingos de Souza Leão entregou a administração da Província ao Exm. Sr. 1º Vice-Presidente Desembargador Anselmo Francisco Peretti* (Recife, 1864), p. 22. *Relatório com que o Exm. Sr. Dr. Manoel do Nascimento Machado Portella passou a administração desta província ao Exm. Sr. Conselheiro João José de Oliveira Junqueira à 27 de outubro último* (Recife, 1872), p. 22.

cial president in 1869 lamented that "discouragement has dominated this institution," and another official blamed

The inertia of vain and incompetent people who only awake at the beckon of a decoration or title, who only busy themselves with problems of municipal judge substitutes, police and National Guard posts, who only become aroused about the creation of judicial districts and the division and creation of parishes for purely political reasons and without respect for the people's comfort or common sense. This presumptuous, incompetent inertia . . . has been more powerful than the will of the founder of these institutes.[38]

The Pernambuco Agriculture Auxiliary Society (SAAP), founded in Recife in 1872, was the second and more enduring agricultural organization. Article 2 of the SAAP statutes stated its goals:

To aid directly the introduction of industrial improvements relative to our farm industry . . . to qualify before the competent authorities to test and realize mutual farm credit associations, to treat agricultural interests with the provincial powers by the available legal means, scrupulously avoiding any government protection other than legal, however beneficent and paternal, so that the society will rely only and always upon its own resources . . . to aid morally its members in their individual commitments or those undertaken at their own expense, which the society judges worthy of this favor, and which are relative to farm improvements, either in the processing of our products or those of large or small scale farming, or in communications and transport, or with relation to the introduction of free labor or animals, or for the better use of the labor and animals of the land.

The SAAP would also perform many of the functions of the moribund IIPA, such as maintain a library and a museum, publish a journal, hold monthly exhibits of products and prices, and keep statistics. To finance these activities, the SAAP would charge individual farmers who benefited from the society, levy assessments, sell its publications and entries to its library and museum, and receive contributions. Sugar planters dominated the SAAP. Eight of the nine Escada sugar oligarchy families par-

38. *Relatório apresentado à Assembléa Legislativa Provincial de Pernambuco pelo Exm. Sr. Conde de Baependy, Presidente da Província na sessão de installação em 10 de Abril de 1869*, p. 20. Paes de Andrade, *Questões Econômicas*, p. 49.

ticipated from the time of its inception, and both the general assembly president and the administrative council chairman invariably were powerful senhores de engenho. A few foreign merchants interested in the sugar trade also belonged.[39]

Finally, the agricultural syndicate movement promoted at the First National Agricultural Congress (1901) led to the organization in 1905 of the Syndicato Agricola Regional de Gamelleira, Amaragy, Bonito e Escada, and in 1906 of the União dos Syndicatos Agricolas de Pernambuco (USAP). The USAP proposed to promote agricultural schools and a laboratory, publish a technical and statistical bulletin, lobby on transport problems, act as a buying and selling agent, distribute seeds, and promote credit cooperatives.[40]

These various class organizations, led by men sympathetic to the sugar interests, articulated the more urgent demands of that sector. Their discussions, conducted at regular meetings, through periodical publications, and at special congresses convoked to consider particular problems, served two purposes. They exchanged views and informed each other of relevant developments, and they lobbied before local and national governments as representatives of the Pernambuco commercial and agricultural interests.

An increasingly acute regional consciousness emerged as one product of the associations' activities. When the Agriculture minister, Alagoas-born João Lins Vieira Cansansão de Sinimbú, convened an agricultural congress in Rio de Janeiro in July 1878, and invited only center-south provinces, northeastern planters were furious. The SAAP soon convened the Congresso Agrícola do Recife in October 1878. This congress, the first of its kind in the northeast, attracted 288 planters and merchants, mostly Pernambucans with representatives from Rio Grande do Norte, Paraíba, Alagoas, and Piauí, who met for one week primarily to discuss sugar problems.

The congress heard some of the earliest northeastern recri-

39. SAAP, Livro de Atas no. 1, "Estatutos," and passim.
40. IAA, Congressos Açucareiros do Brasil, p. 57. Estatutos do Syndicato Agricola Regional de Gamelleira, Amaragy, Bonito e Escada (Pernambuco, 1905). Estatutos da União dos Syndicatos Agricolas de Pernambuco approvado em Assembléia Geral dos Syndicatos a 6 de março de 1906 (Recife, 1906), article 2.

minations against the coffee-producing center-south provinces. In particular, the northeasterners criticized the Banco do Brasil, the government's powerful emission bank, which had loaned 25,000 contos to the center-south provinces of Rio de Janeiro, São Paulo, Espírito Santo, and Minas Gerais in the 1870's. The Pernambuco planters reacted bitterly to this imperial discrimination, exclaiming on the eve of their congress: "There in court, all deference; here in Pernambuco, they throw us favors as one throws food on the ground to starving dogs who have been barking."[41] Others called attention to the differences in living standards between the northeast and center-south. A leading merchant lamented:

Here [Pernambuco], one sees only decency, and one spends only the necessary; here, sumptous dances, excessive luxury and the squandering of the cities are not known; here, one does not see magnificent palacial homes, important parks, beautiful gardens; perhaps many would even have trouble understanding how sugar cane farming can afford a life so full of pleasures, so replete with delights, with so many comforts, surrounded by so much luxury, as that of the happy planters of Quissaman, of Rio de Janeiro, a veritable paradise on earth according to the beautiful descriptions in the newspapers.[42]

And a planter spokesman elaborated this theme:

Thus, while the planter in those provinces, protected by the state, with the advantage of credit, enjoys all comfort and displays an Asiatic luxury, inspecting his plantations while reclining in spacious carriages, wheeling along good roads and solid bridges, we, the planters of the north, are, with exceptions, obliged to restrict ourselves to the most limited subsistence, and to superintend our plantations while mounted on skinny castrated horses or heavy-headed burros, crossing rivers without bridges, roads, mudholes, and ruts, steep hills and fearful precipices, and what is more, at the end of every year, instead of a credit in the correspondente's account, the majority of our fellow farmers are owing them considerable sums.[43]

41. "Agricultura. Reunião agricola em o palacio da presidencia," *Diário de Pernambuco*, August 29, 1878. Numerous angry, bitter references to the Banco do Brasil appear in SAAP, *Trabalhos do Congresso Agrícola*, pp. 27, 64, 134, 139, 183-195, 274, 374.

42. Laurino de Moraes Pinheiro, "Publicações a Pedido: Congresso Agrícola do Recife," *Diário de Pernambuco*, October 16, 1878.

43. S.B., "Bancos Agrícolas," *O Brazil Agrícola*, Ano II, January 31, 1881, pp. 76-77.

When ten years later Pernambuco senator João Alfredo Correia de Oliveira Andrade became president of the Imperial Council of Ministers on March 10, 1888, the opposition press in Recife criticized his failure to act promptly:

The promises to look for intermediaries for the northern provinces are a sad mockery and a scorn thrown at this northern agriculture . . . the government wants to reduce the north to misery.

Let no one be fooled. Mr. Councilor João Alfredo did not give, is not giving, and will not give to the agriculture in the north of the Empire aid equal to that given to the agriculture in the south.[44]

Even after Pernambuco received imperial subsidies, the press insisted:

To what does São Paulo owe in large part its economic expansion? To the great immigration wave which replaced slave labor and more than filled the ranks of rural workers, depleted by abolition. And who but the central government aided this immigration wave with large subsidies? Were not the funds marked for immigration spent mostly in the state of São Paulo? In the system of aid to agriculture created by Mr. João Alfredo, developed by the Visconde de Ouro Preto, and maintained by Mr. Ruy Barbosa, did not São Paulo get the largest share? Meanwhile we, despite the fact that we were the neediest, were working for the benefit of the Brazilian union.[45]

The first ten years of the new Republic witnessed massive state loans to build usinas, but this aid only partly muted the cries of regional jealousy of the late Empire. It did not offset the sugar sector's reverses in foreign markets, and the coffee prosperity annually increased the discrepancy between the northeast and the center-south. The sugar oligarchy found itself in a frustrating contradiction. Within Pernambuco its power, based on control of land, went unchallenged. Outside, the sharper competition in world sugar markets and the burgeoning coffee economy in the center-south reduced its strength. Although the industry had modernized, and still dominated Pernambuco, from a national perspective its importance had diminished.

44. "O norte e o governo," *Jornal do Recife*, June 27, 1888. "O silencio do Governo," *ibid.*, August 4, 1888. For the sugar merchants' ACA's plea to João Alfredo, see "Associação Commercial Agricola," *Diário de Pernambuco*, July 28, 1888.

45. "S. Paulo e o Norte," *Jornal do Recife*, April 6, 1890.

7

SOCIAL CRISIS:
SLAVERY AND GRADUAL
ABOLITION

$\mathscr{S}$lave labor predominated on the sugar plantations in the earlier nineteenth century. The Brazilian government's policy of gradual abolition, however, compelled the planters to abandon the slave system. They attempted in several ways to delay abolition, but they recognized that it was inevitable and in some ways they even helped reduce the slave population, and derived benefits from the transition to free labor.[1]

Slavery

Travelers in the decade before independence reported that planters needed at least 40 able slaves to make sugar; large estates used 100 to 150, and the largest worked as many as 300 blacks. In the early 1840's, the average number of slaves on 331 plantations was 55. A decade later, a survey of 532 plantations reported an average labor force of 20 slaves and 6 free workers; and in 1857, the police chief counted an average of 70 slaves and 49 free individuals, the latter between the ages of 18 and 50, on 46 plantations in Jaboatão, one of the province's richer sugar districts. The slave population constituted between one-fifth and

1. A shorter verison of this and the following chapter appeared in the *Hispanic American Historical Review*, v. 52, no. 4 (November 1972). See footnote 1 of that article for mention of those scholars, such as Emilia Viotti da Costa and Eugene D. Genovese, with whom I substantially agree on the smooth transition from slave to free labor, and also of those writers such as Robert B. Toplin and J. H. Galloway who found violence and even structural change accompanied the abolition process.

one-fourth of the total provincial population around mid-century (Table 22).[2]

Provincial censuses of 1829 and 1842 indicate that between 41 percent and 54 percent of the slave population came from Africa.[3] This high proportion reflects the fact that the planters did not replenish their slave labor force through natural reproduction but rather through constant imports from Africa. During the years 1839-50, for example, Pernambuco annually imported at least 1,100 slaves, and in some years as many as 3,000 (Table 23).

2. Koster, *Travels in Brazil*, v. II, pp. 138-139, 218-223. Tollenare, *Notas Dominicais*, pp. 71, 74, 93. Figueira de Mello, *Ensaio sôbre a Statística*, p. 263. *Relatório . . . 1854 . . . José Bento da Cunha e Figueiredo*, table. "Uma Estatistica," *Diário de Pernambuco*, December 15, 1857.

3. Figueira de Mello, *Ensaio sôbre a Statística*, pp. 202, 208.

TABLE 22

PERNAMBUCO POPULATION

Year	Slaves	Free	Total	Slaves/Total (percent)
1772-82			239,713	
1775			245,000	
1810			274,687	
1814			294,973	
1815			339,778	
1819	97,633	273,832	371,465	35.7
1823	150,000	330,000	480,000	31.3
1829	80,265	208,832	287,140	28.0
1832			550,000	
1839	146,500	473,500	620,000	23.6
1842	146,398	498,526	644,924	22.7
1855	145,000	548,450	693,450	20.9
1872	89,028	752,511	841,539	10.6
1873	106,236			
1882	84,700			
1883	83,835			
1886	80,338			
1887	41,122			

(Continued)

148   THE SUGAR INDUSTRY IN PERNAMBUCO

SOURCES: *1772-82:* Dauril Alden, "The Population of Brazil in the late Eighteenth Century: A Preliminary Survey," *Hispanic American Historical Review,* v. XLIII, no. 2 (May 1963), p. 191.

1775, 1814, 1819: Souza e Silva, "Investigações sôbre os recenseamentos," p. 47.

1810, 1815, 1829, 1839, 1842: Figueira de Mello, *Ensaio sôbre a Statística,* pp. 160, 163, 164, 202, 208. We have combined the proportion of slaves in the total counted populations in 1839, and 1842, with the total estimated population for these years.

1823, 1832: Oliveira Vianna, "Resumo Histórico dos Inquéritos Censitários," pp. 405, 462.

1855: Cowper to Clarendon, Pernambuco, July 18, 1855, in *Parliamentary Papers, 1856,* HCC, v. LXII, *AP,* v. XXX, p. 239.

1872: Recenseamento da população . . . *1872,* v. XIII, p. 215.

1873, 1882: Vieira Souto cited by Rui Barbosa in Padua, "Um Capítulo da História Econômica," p. 163. We should note here that the 1873 figure, which represents the total number of slaves entered in the special register created by the Law 2,040 of 1871, is more reliable than the 1872 figure, which represents the total number of slaves counted in the first official national census. According to the 1871 law, slaves not entered in the special register within one year after its opening would be considered free. The difference between the 1873 and the 1872 counts represents those slaves not enumerated in the census, but hastily registered by their owners subsequently.

1883: "Elemento servil," *Diário de Pernambuco,* March 27, 1885.

1886: Relatório com que o Sr. Francisco Augusto Pereira da Costa dá conta ao Exm. Sr. Presidente da Provincia da Commissão de que fora encarregado em 2 de março de 1866, p. 50, cited in Galloway, "The Sugar Industry of Pernambuco," p. 298, gives 80,872 for 1886. "População Escrava," *Diário de Pernambuco,* January 29, 1886, gives 79,803 for 1886. Our figure is the average of the two numbers.

1887: Relatório . . . *Agricultura, Commercio e Obras Publicas Rodrigo Augusto da Silva* (1888), p. 24.

Natural reproduction did not satisfy the demand for slaves for three reasons related to working conditions. First, the strenuous labor on sugar plantations led planters to prefer the stronger male laborers; this preference skewed the sex distribution among African imports in favor of males, and the resulting relative scarcity of slave women inhibited reproduction (Table 24). Secondly slave women tended to bear children about half as frequently as free women, as indicated by fertility ratios for 1829 and 1842: this relative infertility was attributed by nineteenth-century observers to insufficient care for pregnant women and newborn infants. Even when the African slave trade stopped and children born of slave mothers were declared free, slave fertility remained low. Thus records kept on slaves' children after 1871 show that annual births totaled 2,300, or 30.6 per thousand in an average slave population of 73,679 between 1873 and 1887. Since general mortality never decreased below 27 deaths per thousand in Recife, usu-

TABLE 23

PERNAMBUCO SLAVE IMPORTS

Year	Yearly Totals	Period	Estimated Totals for Periods	Annual Averages Based on Estimated Totals
1804	3,325 [a]	1801-23	130,418	5,670
1805	1,401			
1809	2,494	1801-39	150,000	3,846
1810	1,254			
1812	2,489	1817-43	6,600 [b]	244
1813	3,265			
1815	3,911	1831-43	20,000	1,539
1816	5,499			
1817	5,932	1840-43	12,000	3,000
1818	7,702			
1819	7,802 [c]	1839-50	12,512	1,043
1823	4,824 [a]			
1824	2,683			
1825	2,384			
1847	300			
1849	450			
1850	2,300			
1855	240			

[a] Figures for 1804-18 and 1823-25 include only slaves from Angola.

[b] Includes only slaves counted by British oficials.

[c] Includes slaves from Benguela.

SOURCES: *Yearly Totals*. *1804-25*: Edmundo Correia Lopes, *A Escravatura (subsídios para a sua história)* (Lisbon, 1944), pp. 141-142. *1847-50*: Christophers to Palmerston, Pernambuco, March 12, 1851, *Parliamentary Papers*, 1852-53, HCC, v. CIII, *AP*, v. XLVII, part II, p. 379. *1855*: Bethell, *The Abolition of the Brazilian Slave Trade*, p. 395.

SOURCES: *Estimated Totals*. *1801-23*: *Rezumo da Importação da Provincia de Pernambuco* (1823), cited in Affonso de Escragnolle Taunay, *História do Café no Brasil*, 15 vols. (Rio de Janeiro, 1939-43), v. 4, *No Brasil Imperial, 1822-72*, Tomo II, p. 228. Vicente Cardoso, *A margem da história do Brasil* (São Paulo, 1933), p. 171, also cites Alcides Bezerra, who showed Taunay the *Rezumo*, but Cardoso states that between 1800 and 1817 alone, 130,000 slaves were imported. *1801-39*: Mauricio Goulart, *Escravidão Africana no Brazil (Das origens à extinção do tráfico)* (São Paulo, 1949), p. 272. *1817-43*: Philip D. Curtin, *The Atlantic Slave Trade. A Census* (Madison, Wisconsin, 1969), p. 240. *1831-43, 1840-43*: Cowper to Aberdeen, August 4, 1843, *Parliamentary Papers*, 1844, HCC, v. XLIX, *AP*, v. XVIII, pp. 368-369, 383. *1839-50*: Galloway, "The Sugar Industry of Pernambuco," p. 297.

TABLE 24

VITAL INDEXES FOR PERNAMBUCO SLAVES AND TOTAL POPULATION

Index	Year					
	1829	1842	1872	1887	1890
Sex Ratio (males/100 females)					
Slaves: Africans	162	156	138		
Brazilians	110	110	111		
Total	136	133	112	100	
Total Population	104	100	104		96
Fertility (children under age 10/1,000 Women ages 11-50)					
Slaves	439	640	868		
Total Population	756	1,010	866 [a]		
Median Age					
Slaves	21-30	21-30	21-25	31-40	
Total Population	21-30	11-20	26-30 [a]		18-19

[a] The 1872 census underenumerated the free young, with the result that the fertility index of the total is understated and the median age of the total is exaggerated. This error is common to the national census as well. See Eduardo E. Arriaga, *New Life Tables for Latin American Populations in the Nineteenth and Twentieth Centuries* (Berkeley, 1968), pp. 25-26.

SOURCES: Figueira de Mello, *Ensaio sôbre a Statística*, pp. 202, 208. *Recenseamento da População . . . 1872*, v. XIII, pp. 214-215, 217. *Relatorio Apresentado à Assembléa Geral na Terceira sessão da vigéssima legislatura pelo Ministro e Secretário de Estado dos Negocios da Agricultura, Commercio e Obras Publicas Rodrigo Augusto da Silva* (Rio de Janeiro, 1888), p. 24. Directoria Geral de Estatística, *Sexo, raça e estado civil, nacionalidade, filiação culto e analphabetismo da população recenseada em Dezembro de 1890* (Rio de Janeiro, 1898), pp. 94-99. Idem, *Idades da População recenseada em 31 de dezembro de 1890* (Rio de Janeiro, 1901), pp. 276-291.

ally ranging between 29 and 35 per thousand, and slaves notoriously died more frequently than free people, the slaves must have suffered a negative rate of natural increase. Lastly, young slaves suffered heavy mortality. Of the 36,807 children born of slave mothers and registered between 1871 and 1887, 8,545 died, an astounding age-specific mortality rate of 232 per thousand for the cohort 0 to 16 years old. This rate more than doubles the mortality recorded for the free cohort 0 to 20 years old in Recife between 1894 and 1904, and is 27 percent above the highest mortality rate reported for any age cohort during those eleven years.[4]

Gradual Abolition

As long as planters could depend upon fresh slave imports from Africa, the failure of natural reproduction posed no insoluble problems. Thus the absolute numbers of slaves increased in the early nineteenth century, even though their importance in the total population was declining. The proportional decline probably resulted from interruptions in slave imports. Between 1811 and 1815, the Napoleonic Wars effectively restricted the volume, if not the value, of Pernambuco sugar exports, which fell 27 percent below the previous five-year average. In the second half of 1824, the provincial revolt known as the "Confederation of the Equator" caused disruptions which cut sugar production in 1825 and 1826 to below half the 1824 level. Similarly, the prolonged "Cabanada" uprising between 1832 and 1836, so disturbed sugar trade that

4. *Relatório da Commissão dirigida por Joaquim d'Aquino Fonseca, apresentado ao Excellentissimo Sr. Conselheiro Dr. José Bento da Cunha e Figueiredo em 10 de janeiro de 1856* (Pernambuco, 1856), pp. 28-30. An English plantation manager reported that only on estates run by monks, and on another directed by three old women and a priest, did natural reproduction maintain the labor force. Presumably these plantation owners were more considerate with their slaves. Koster, *Travels in Brazil*, I, 258; II, 217-222. *Relatório com que o Excellentissimo 1º Vice-Presidente Dr. Ignácio Joaquim de Souza Leão passou a administração da Provincia em 16 de abril de 1888 ao Excellentissimo Presidente Desembargador Joaquim José de Oliveira Andrade*, p. 19. Our calculation of average population is the mean of figures in Table I. Octavio de Freitas, *O Clima e a Mortalidade da Cidade do Recife* (Recife, 1905), pp. 62-63. Tollenare, *Notas Dominicais*, p. 76, had estimated a negative rate of natural increase between 2 and 5 percent.

152   THE SUGAR INDUSTRY IN PERNAMBUCO

the 1831-35 production average fell 25 percent below that of the preceeding five-year period (Table 5). These internal disruptions not only reduced production, but also, by shrinking demand for productive factors, reduced the slave import trade. Since the free population continued increasing naturally, despite the casualties of revolts, the proportion of slaves declined.[5]

When the African slave supply stopped completely, however, the proportional decline accelerated and the absolute numbers of slaves began to fall. As early as November 7, 1831, the Brazilian government had ordered the liberation of all slaves thereafter entering Brazil. For a variety of reasons, however, the 1831 law had remained largely unenforced, and Pernambucans continued to import African slaves, although in reduced numbers. Then in 1845, with an eye on enlarging foreign markets as well as promoting humanitarian ideals, the British Parliament passed the Aberdeen Bill authorizing British courts to deal with Brazilian slave traders captured by the British Navy. To this new threat, the Brazilians reacted by increasing slave imports, reportedly in patriotic resentment against British interference and in response to increased foreign demand for Brazilian coffee and sugar.[6]

But such defiance did not last long. To preserve the national honor against the unreturnable insult of British men-of-war sailing into Brazilian harbors to arrest slave traders, the Brazilian Emperor approved Law No. 581, also known as the Eusebio de Queiróz Law, on September 4, 1850. This law and its imple-

5. For summaries of the revolts, see Amaro Quintas, "A Agitação Republicana no Nordeste," Buarque de Holanda, *História Geral da Civilização Brasileira*, tomo II, v. 1, pp. 207-237; and *ibid.*, v. 2, "O Nordeste, 1825-1850," pp. 193-311. Another revolt in 1817, despite the fact that the combating armies moved through the sugar zone, did not apparently affect sugar production. In fact, slave imports increased noticeably between 1816 and 1819, allegedly due to a small cotton boom. Correia Lopes, *A Escravatura*, p. 141.

6. Paula Beiguelman, "O encaminhamento político do problema da escravidão no império," Buarque de Holanda, *História Geral da Civilização Brasileira*, tomo II, v. 3, pp. 189-201. Robert Conrad, "The Contraband Slave Trade to Brazil, 1831-1845," *Hispanic American Historical Review*, v. XLIX, no. 4 (November 1969), pp. 617-638. Bethell, *The Abolition of the Brazilian Slave Trade*, pp. 62-213, 284. João Pandiá Calógeras, *A History of Brazil*, translated and edited by Percy Alvin Martin (Chapel Hill, N.C., 1939), p. 189.

menting regulations stipulated penalties for infractions of the earlier 1831 law; implicitly backed by the British Navy, Law No. 581 won general compliance.[7]

The effective abolition of the international slave traffic initiated the gradual abolition process by preventing the renovation of one-half the slave labor force. The restricted supply caused nominal slave prices in Pernambuco to more than triple by 1860 (Table 25).

7. "Lei 581," *Colleção das Leis do Império do Brazil de 1850*, Tomo XI, Parte I, pp. 203-205. E. Bradford Burns (editor), *A Documentary History of Brazil* (New York, 1966), provides an easily accessible English translation of this and other laws of gradual abolition. Bethell, *The Abolition of the Brazilian Slave Trade*, pp. 327-341, reviews the debates preceding passage of the law.

TABLE 25

PERNAMBUCO SLAVE PRICES, 1852-1887

(average prices in mil-réis for males and females, 20-25 years old)

Year	Nominal Price	Real Price (1852=100)	Year	Nominal Price	Real Price
1852	450	450	1871	767	441
1853	503	347	1872	650	387
1854	688	430	1874	456	285
1857	1:200	563	1875	400	204
1858	1:467	539	1876	670	390
1859	1:139	458	1877	644	280
1860	1:500	735	1878	698	255
1861	1:243	730	1879	886	372
1862	867	628	1880	683	382
1863	1:158	816	1881	578	274
1864	800	684	1882	350	151
1866	775	615	1883	750	397
1867	667	379	1884	590	306
1868	700	440	1887	283	181
1870	1:425	848			

SOURCE: Inventories included in wills in the Cartório Público de Ipojuca, Pernambuco. A total of 212 slave prices were found for the 29 years listed, a median number of 4 prices per year. The nominal values were deflated by a cost of living index based on 1852 prices (Table 26).

TABLE 26
PERNAMBUCO PRICE INDEXES [a]
Index A 1852-88

Year	Nominal Value (réis) [b]	Index Number	Year	Nominal Value (réis)	Index Number
1852	$053	100	1870	$089	168
1853	$077	145	1871	$092	174
1854	$085	160	1872	$089	168
1855	$093	176	1874	$085	160
1856	$117	221	1875	$104	196
1857	$113	213	1876	$091	172
1858	$144	272	1877	$122	230
1859	$132	249	1878	$145	274
1860	$108	204	1879	$126	238
1861	$098	185	1880	$095	179
1862	$073	138	1881	$112	211
1863	$075	142	1882	$123	232
1864	$062	117	1883	$100	189
1865	$064	121	1884	$102	193
1866	$067	126	1885	$099	187
1867	$093	176	1886	$083	157
1868	$084	159	1887	$083	157
1869	$087	164	1888	$071	134

Index B 1886-1903 [c]

Year	Nominal Value	Index Number	Extrapolated Index Number (1852=100) [d]
1886	$282	100	157
1887	$208	74	116
1888	$278	99	157
1889	$422	150	236
1890	$374	133	209
1891	$261	93	146
1892	$432	153	240
1893	$425	151	237
1894	$636	226	355

Year	Nominal Value (réis)	Index Number	Extrapolated Index Number (1852=100)
1895	$760	270	424
1896	$646	229	360
1897	$638	226	355
1898	$926	328	515
1899	$815	289	454
1900	$680	241	378
1901	$543	193	303
1902	$430	153	240
1903	$368	131	206

ᵃ Discontinuities in price data obliged us to construct two different indexes. The overlap in the years 1886-88 allows extrapolation to calculate values in the period following 1888 in terms of 1852 prices; this is given in Index B. Price data comes from the weekly commodity price listing in the *Diário de Pernambuco*. We averaged six prices per year, two each from the months of January, May, and September for the first five years of each decade (for example, 1860-64) and two each from the months of March, July, and November for the fifth through ninth years of each decade (for example 1865-69). Conversions to metric wrights followed equivalents listed in our Table of Measures.

ᵇ Values in Index A reflect wholesale costs of 400 grams of manioc flour and 200 grams of dried or jerked meat, which together with 125 grams of beans constituted the minimum daily diet in the period. Bean prices were not available for the whole period. See *Relatório da Comissão Central de Socorros aos indigentes victimas da secca* (Pernambuco, 1878), p. 5. Earlier Tollenare had found that slaves were fed 455 grams of manioc and 199 grams of jerked·meat daily; Tollenare, *Notas Dominicais*, p. 75. In the twentieth century, various authors confirm the importance of jerked meat, manioc, and beans in rural workers' diets. Josué de Castro, *Documentário do Nordeste*, 3rd ed. (São Paulo, 1965, first published 1937), pp. 70, 74. Dé Carli, *Aspectos Açucareiros de Pernambuco*, pp. 19-20. Vasconcelos Torres, *Condições de Vida do trabalhador no agro-industria do açucar* (Rio de Janeiro, 1945) pp. 181, 201, 213. The proportional weights of the foods in the daily diets varied.

ᶜ Values in Index B reflect wholesale costs of 400 grams of manioc flour and 125 grams of beans.

ᵈ The use of different commodities in the two indexes allowed certain discontinuities in our extrapolation—for example, Index A remained stable in 1887 and fell in 1888, while extrapolated Index B, also in 1852 prices, fell in 1887 and rose in 1888. The Pernambucan historian Gadiel Perucci has developed a more detailed price index for the period after 1889. Perucci, "Le Pernambouc (1889-1930)," chapter IV.

Part of this increase can be ascribed to the inflation of the
later 1850's, a result of increased emissions by banks often funded
by slave trade capital. But even in terms of 1852 prices, the value
of slaves increased by 50 percent by 1860, and nearly doubled by
1870. In coffee-producing Rio de Janeiro, moreover, nominal
slave prices rose even higher, and reached a peak in the late
1870's at a level nearly four times that of the early 1850's. The coffee sector's greater prosperity allowed the coffee planters to out-
bid the sugar planters for slaves, and after 1850 Pernambuco
began shipping slaves south.[8]

This interprovincial slave trade flourished for three decades,
1850 to 1880. The sugar planters sold their slaves in small lots of
a few each year to cover debts held by their factors in Recife, and
an annual average of 760 slaves legally left the province (Table
27). Because the slave owner had to pay an exit tax of 100$000
per slave after 1852, and 200$000 after 1859, many slaves were
smuggled south. Thus the actual total shipped south was probably between 1,000 and 1,500 each year. Most of these slaves
were "young, highly productive men."[9] The interprovincial
slave trade reached it's peak in the late 1870's as a result of severe

8. Calógeras, *A Política Monetária do Brasil*, chapters VII-VIII. Stein,
Vassouras, pp. 52, 229. Stein does not calculate deflated slave prices. Our data
contradict J. H. Galloway, "The Last Years of Slavery on the Sugar Plantations of
Northeastern Brazil," *Hispanic American Historical Review*, v. 51, no. 4 (November 1971), p. 590.

9. "Informe" by Henrique Augusto Milet in *Falla. . . Adolpho de Barros Cav^{te} de Lacerda. . . 19 de Dezembro de 1878*, p. 30. Ferreira Soares,
Notas Estatísticas, pp. 135-136, calculated that the central and southern provinces of Rio de Janeiro, Minas Gerais, and Rio Grande do Sul alone imported a
total of 5,500 slaves per year from the north. If Pernambuco's share of this total
equaled its share of the total decline in slave population in the northern and
northeastern provinces between 1873 and 1882, then 18.5 percent of Ferreira
Soares' total, or 1,020 slaves, would have left Pernambuco annually. Stein, *Vassouras*, p. 295. A British consul in Recife estimated a total of 1,500 slaves
shipped south each year. Cowper to Clarendon, Pernambuco, January 24, 1857,
Parliamentary Papers, 1857, HCC, v. XLIV, *AP*, v. XX, p. 261. Quote from
Robert Conrad, *The Destruction of Brazilian Slavery, 1850-1888* (Berkeley and
Los Angeles, 1972), p. 62. Conrad believes that the interprovincial trade also
led masters to discourage slave marriages in the northeast. *Ibid.*, pp. 32-33.

droughts in the northeast, which forced liquidation of fixed assets such as slaves. The volume of slaves shipped south after 1876 was so high that the major slave-buying provinces of Rio de Janeiro, São Paulo, and Minas Gerais imposed prohibitively high entrance taxes on slaves in 1880 and 1881. These taxes were levied to prevent draining all slaves from the northeast and thereby disposing those provinces to support abolition, and also to encourage European immigration. The taxes ended the interprovincial slave trade. As a result of this trade, Pernambuco may have lost between 23,000 and 38,000 slaves, depending upon whether one uses the legal annual average shipment or this estimated average shipment.[10]

Two other laws of gradual abolition also limited the slave population. Law No. 2,040 of September 28, 1871, popularly known as the Rio Branco or Free Womb Law, freed children thereafter born of slave mothers, with certain qualifications. The children would remain wards of their mother's master until age eight, when the master had two options. He could free the eight-year-old and receive state indemnification, or he could keep the child until age 21, when freedom would be granted without indemnifi-

10. Report by Walker, May 29, 1878, *ibid.*, 1878, HCC, v. LXXV, *AP*, v. XXX, p. 10. Emilia Viotti da Costa, *Da Senzala à Colônia* (São Paulo, 1966), pp. 208-209. Padua, "Um Capítulo da História Econômica," pp. 180-182. Gilberto Freyre mistakenly dates the interprovincial slave trade after 1880. *Order and Progress. Brazil from Monarchy to Republic*, edited and translated by Rod W. Horton (New York, 1970, first published as *Ordem e Progresso*, 1959), p. 229. The 1872 census counted 6,820 slaves born in Pernambuco but living in provinces to the south. If one knew the life expectancies of these slaves, one could calculate all slaves shipped south from Pernambuco since 1850, and check our estimate. But the group was not identified by age. *Recenseamento da população . . . 1872*, vv. 1-21, *passim*. Herbert S. Klein implies the possibility that the interprovincial trade continued despite taxes and laws until 1888. If so, then the totals exported from Pernambuco would range between 38,000 and 57,000. 'The Internal Slave Trade in Nineteenth-Century Brazil: A Study of Slave Importations into Rio de Janeiro in 1852," *Hispanic American Historial Review*, v. 51, no. 4 (November 1971), p. 569. For a description of coffee-growers' pressures to end the interprovincial slave trade, see Robert Brent Toplin, *The Abolition of Slavery in Brazil* (New York, 1972), pp. 90-91.

TABLE 27

SLAVES LEAVING PERNAMBUCO

Year	Legal Exits	Year	Legal Exits
1851[a]	270	1861	1,016
1852[a]	123	1862	516
1855	606	1863	1,033
1856	145	1864	285
		1872-75 [b]	1,753
1857	5	1876	1,300
1858	2	1877	1,271
1859	139	1878	2,212
1860	950	1879	1,329

[a] Figures for first semester only.

[b] Annual average.

SOURCES: *1851-52: Relatório que à Assembléa Legislativa Provincial de Pernambuco apresentou na abertura da sessão ordinária em 1 de março de 1853 o Excellentíssimo Presidente da Província Francisco Antônio Ribeiro*, p. 4. *1856-65: Relatório que o Illustrissimo e Excellentissimo Senhor Dr. Antonio Borges Leal Castello Branco apresentou ao Illustrissimo e Excellentissimo Senhor Conselheiro João Lustosa da Cunha Paranaguá tendo entregado a administração da província ao Illustrissimo e Excellentissimo Senhor Barão do Rio Formoso* (Recife, 1865), p. 60. *1872-75:* Directoria Geral de Estatistica, *Relatório e Trabalhos Estatisticos apresentados ao Illm. e Exm. Sr. Conselheiro Dr. Carlos Leoncio de Carvalho Ministro e Secretário de Estado dos Negócios do Império pelo Director Geral Conselheiro Manoel Francisco Correia em 20 de Novembro de 1878* (Rio de Janeiro, 1879) p. 129. We deduct 1,300 of 1876 from the total of 8,311 for the period 1872-76. *1876-79:* Report by Bonham, Pernambuco, April 30, 1881, *Parliamentary Papers*, 1881, HCC, v. XCI, *AP*, v. XXXV, p. 112.

cation. The law also created an emancipation fund based on tax revenues, lotteries, and fines.[11]

Between October 1871 and the end of 1887, a total of 37,000 children of slave mothers were reported to the provincial president. Probably few of these children enjoyed legal freedom, however, for planters generally refused the small indemnity and kept the children working for them after age eight. Other slave owners simply abandoned the children, who could not be legally sold. The Imperial Emancipation Fund freed 2,600 young slaves by

11. "Lei 2,040," *Colleção das Leis do Imperio do Brasil de 1871*, Tomo XXXI, Parte I, pp. 147-151.

May 1888. A Provincial Emancipation Fund created in May 1883 freed 150 slaves.[12]

Law No. 3,270, the Dantas-Saraiva-Cotegipe or Sexagenarian Law of September 28, 1885, liberated all slaves aged 60 or older. Like the Free Womb Law passed exactly fifteen years previously, however, the Sexagenarian Law qualified even its mild provisions by obliging the elderly slaves to continue working for their former masters for another three years, or until they reached age 65. The 1885 law also granted a subsidy to the Imperial Emancipation Fund and formally outlawed the interprovincial slave trade, which had already been practically extinguished. The Sexagenarian Law freed no more than 9,600 slaves in Pernambuco.[13]

The imperial legislature passed the Golden Law of Abolition, Law No. 3,353, on May 13, 1888, and freed all remaining slaves in the Empire. A few Pernambuco towns and parishes had anticipated the Golden Law by freeing slaves within their jurisdiction in the early months of 1888. In the last count prior to abolition, in 1887, the Pernambuco slave population was reported to be near 41,000.[14]

12. *Relatório . . . Ignácio Joaquim de Souza Leão . . . 16 de abril de 1888*, p. 19. "Já é tempo," *O Brazil Agricola*, Anno IV, April 30, 1882, p. 121. "Breves considerações sobre a Agricultura no Brasil, IX," *Diário de Pernambuco*, April 27, 1876. "Libertos pelo fundo de emancipaço," *ibid.*, March 23, 1888, reported 2,579 young slaves freed by December 1887; we have projected this rate until May 1888 to arrive at our total. *Annaes da Assembléa Provincial de Pernambuco do anno de 1883*, p. 21. *Falla que à Assembléa Legislativa Provincial de Pernambuco no dia de sua installação a 2 de março de 1887 dirigio o Excellentíssimo Sr. Presidente da Provincia Dr. Pedro Vicente de Azevedo*, p. 79, reported 105 sexagenarians freed; we have projected this rate into the remaining months before abolition.

13. "Lei 3,270," *Colleção de Leis do Imperio de Brazil do 1885*, Tomo XXII, Parte I, pp. 14-20. "Libertos Sexagenarios," *Diário de Pernambuco*, September 20, 1887. We calculate the sexagenarians freed in 1887 and early 1888 by presuming that all slaves counted between ages 44 and 45 in 1872—that is, 15 percent of the presumably evenly distributed total age 41-50 cohort—survived until 1888, when they would have passed age 60. The presumptions are obviously unlikely, and the total freed by the Sexagenarian Law was therefore really less than 9,600. Moreover, the law was widely evaded. Toplin, *The Abolition of Slavery in Brazil*, pp. 108-109.

14. "Lei 3,353," *Colleção de Leis do Império do Brazil de 1888*, Tomo XXV, Parte I, v. I, p. 1.

In addition to legal measures, epidemic disease and other causes of death also reduced the slave population; in fact, deaths depleted that population far more than any other cause. The worst cholera epidemic of the century ravaged Pernambuco in 1855-56, nearly tripled the normal death rate in Recife of 29 deaths per thousand, and killed at least 3,300 individuals, including "a vast quantity of slaves." Yellow fever epidemics raged in the early 1860's, along with cholera, and killed many slaves, "that portion of the population whose habits and mode of life rendered them the earliest victims." Yellow fever reappeared in 1871 and 1873, when the urban death rate ranged between 37 and 41 per thousand. Smallpox struck during the drought years of 1878 and 1879 and killed 2,500 in Recife; in the late 1880's it again killed 2,200. Without complete vital statistics or details on the age and sex of slaves leaving the province or manumitted, however, one cannot precisely quantify the net impact of all causes of death on the slave population. Certainly slaves died at least as frequently as free people. Thus in the average slave population between 1850 and 1873 of 126,368, the minimum Recife death rate of 27 per thousand would have taken at least 3,413 slave lives annually. Between 1871 and 1888, the average slave population of 73,679 would have suffered a minimum loss of 1,989 annually.[15]

Manumissions and emancipations removed individuals from the slave population, and represented the only major pressure on that population before 1888 which may have resulted in a change beneficial to the slaves. According to Brazilian law, slaveowners were required to grant their slaves freedom under certain circumstances. These laws for formal manumission have been summarized by several modern writers and need not detain us here. It is doubtful that these laws were regularly enforced: at least one competent nineteenth-century observer specifically stated that he found infrequent compliance in Pernambuco.[16]

15. Freitas, *O Clima e a Mortalidade*, pp. 56, 82-83. Quotes from Cowper to Clarendon, Pernambuco, December 19, 1855, *Parliamentary Papers*, 1856, HCC, v. LXII, *AP*, v. XXV, p. 245; and Report by Hunt, August 18, 1864, *Parliamentary Papers*, 1865, HCC, v. LIII, *AP*, v. XXIV, p. 348.
16. Frank Tannenbaum, *Slave and Citizen, the Negro in the Americas*

Beyond their legal obligations, sugar planters freed certain favored slaves on important occasions such as birthdays, weddings, anniversaries, and deaths. Emancipation did not always reflect generous sentiments; by freeing sick and aged slaves, for example, their owner could eliminate maintenance costs. Especially in the last years of slavery, as the slave population grew older and abolitionism gained favor, emancipations occurred more frequently and gained wider publicity. A total of 6,800 private emancipations were reported between October 1873 and June 1886, an annual average near 600. In the years immediately preceding abolition, the emancipation rate rose sharply: the *Diário de Pernambuco* reported 700 emancipated in the first four and one-half months of 1888.[17]

As in the center-south provinces, slaves freed themselves by fleeing the plantations in increasingly large numbers in the last years of slavery. The fugitives were aided by the abolitionists who often encouraged them to run away, and by the police, who stood by and did nothing to impede them. But the sharp decline in reported slave population between 1886 and 1887 should not be attributed to such escapes, but rather to inaccuracies in reported figures. The provincial governments received quotas in the

---

(New York, 1946), pp. 91ff. Stanley Elkins, *Slavery: A Problem in American Institutional and Intellectual Life* (Chicago, 1959), pp. 21ff. David Brion Davis, *The Problem of Slavery in Western Culture* (Ithaca, N.Y., 1966), pp. 222ff. Koster, *Travels in Brazil*, v. II, pp. 192, 195, 197. An examination of court records for cases based on the manumission laws would permit checking these impressions; time did not allow me to make such a check. Carl N. Degler, *Neither Black nor White. Slavery and Race Relations in Brazil and the United States* (New York, 1971), pp. 40-42, notes that manumission in Brazil was often optional and revocable.

17. Koster, *Travels in Brazil*, v. I, pp. 194, 217, 297. Freyre, *The Master and the Slaves*, pp. 438-439. A visiting American scientist in the later 1870's noted that the "emancipation spirit is very strong" in Pernambuco. Smith, *Brazil, the Amazons, and the Coast*, p. 470. "População escrava de Pernambuco," *Diário de Pernambuco*, January 29, 1886. The plausibility of the annual average is suggested by a brief calculation. The richest sugar planters and their wives composed an oligarchy owning 15 percent or 225 of the province's sugar plantations in 1889. If these oligarchs could afford to free three slaves per year—one for each of the owners' birthdays and another on their wedding anniversary, then these planters alone would be freeing 675 slaves per year.

Imperial Emancipation Fund on the basis of reported slave population: hence exaggerating the number of slaves in the early 1880's would permit the authorities to reimburse more slave owners more generously. Moreover, the only accurate slave count after 1873 was taken in 1887, the second and last slave register. Slave population estimates between 1873 and 1887 depended upon simple deductions of slave deaths and legal exits from the total registered in 1873. The innaccuracies of these figures exaggerated the actual slave population in the early 1880's.[18]

Private abolitionist societies played a considerable but as yet unstudied role in freeing Pernambuco slaves. The first such group, the Associação de Socorros Mutuos e Lenta Emancipação dos Captivos was founded in 1859 by the Bishop João da Purificação Marques Perdigão in Recife. The next year a group of students founded the Associação Acadêmica Promotora da Remissão dos Captivos, and in 1867 a group of Bahian students at the Recife law school formed their own group, the Sociedade Patriótica 2 de Julho. None of these groups survived very long; but in 1872 at least five other abolitionist societies were active in Recife, and by 1884 a total of 21 such organizations, each with over 25 members, operated in the provincial capital, and at least four other groups were active in the interior. The Commissão Central Emancipadora do Recife attempted to coordinate these societies' efforts.[19]

18. "Sociedade Auxiliadora da Agricultura, Segundo Congresso," *Diário de Pernambuco*, June 27, 1884. "Segundo Congresso do Recife. Acta da Primeira Sessão," *ibid.*, July 26, 1884. Robert Brent Toplin has emphasized the importance of slave escapes in precipitating abolition. "Upheaval, Violence, and the Abolition of Slavery in Brazil: the Case of São Paulo," *Hispanic American Historical Review*, v. XLIX, no. 4 (November 1969), pp. 643-644. Decree no. 5,135, November 13, 1872, in Luiz Francisco da Veiga, *Livro do Estado Servil e Respectiva Libertação contendo a Lei de 28 de Setembro de 1871 e os decretos e avisos expedidos pelos Ministerios da Agricultura, Fazenda, Justiça, Imperio e Guerra desde aquella data até 31 de Dezembro de 1875* (Rio de Janeiro, 1876), p. 52. I am indebted to Robert Conrad for this reference, and to Robert Slenes for explaining the pre-1887 estimates.

19. The only summaries of abolitionist societies in Pernambuco were found in Medeiros, "O Movimento da Abolição no Nordeste," pp. 92-96, and Pereira da Costa, *Anais Pernambucanos*, v. X, pp. 423-430. Evaristo de Moraes, *A campanha abolicionista (1879-1888)*, (Rio de Janeiro, 1924), pp. 228-231 concentrated on prominent abolitionists such as José Mariano and José

Most of the abolitionist societies restricted their activities to public meetings in favor of the cause; they bought emancipations with the proceeds of these meetings and other contributions. A few abolitionists took stronger action. In 1876, an abolitionist in the sugar county of Pau d'Alho "penetrated the engenhos, visited the slaves' quarters, advised flight, and promoted revolts by the slaves against the masters."[20] A group founded in 1884, the Sociedade Relâmpago, later renamed the Club Cupim, dedicated itself to "the freeing of slaves by all means." The Club Cupim members arranged emancipations with the slave owners' consent; but they also kept their identities hidden behind *noms de guerre*, held secret meetings, and infiltrated slave quarters to incite escapes. They maintained an "underground railroad," aiding fugitive slaves to hide, to disguise themselves, and to make their way by boat to Ceará, where slavery had been abolished in some counties since 1883 and in the province on March 25, 1884. The Club often sent twenty slaves in a single shipment to Ceará; the last group smuggled out, in April 1888, had 119 slaves.[21]

Over the period 1850-88 as many as 21,000 slaves were freed by manumission, emancipation, and abolitionist activities, if the annual average for 1873-86 can be generalized. But one should note that some 40 percent of those liberations entailed obligations on the part of the ex-slave, such as that he continue "offering services" for two or three years thereafter or that he pay

---

Maria. 'Commissão central emancipadora," *Diário de Pernambuco*, July 3-4, 1884. For some aspects of abolitionism in Pernambuco, see articles listed by José Honório Rodrigues, *Índice Anotado da Revista do Instituto Arqueológico e Geografico Pernambucano* (Recife, 1961), pp. 25-26.

20. Mario Melo, *Pau d'Alho; Geographia physica e política* (Recife, 1918), p. 27.

21. Medeiros, "O Movimento da Abolição no Nordeste," p. 93. See Eusébio de Sousa, "O Quadro Histórico," in Raimundo Girao, *A Abolição no Ceará* (Forteleza, 1956), pp. 253, 265. Rollie Poppino, *Brazil, the Land and the People*, p. 175, errs by dating Cearense abolition as May 24, 1883. Joaquim Maria Carneiro Vilela, "O Clube do Cupim," *Revista do Instituto Arqueológico, Histórico e Geográfico Pernambucano*, v. XXVII, nos. 127-130 (1925-26), pp. 417-427. For minutes of the Club Cupim Meetings, see *Catálogo da Exposição realizada no Teatro Santa Izabel de 13 à 31 de maio de 1938* (Recife, 1939), pp. 31. *Cupim* are termites.

his owner his current value.[22] Therefore, probably many less than 21,000 actually gained freedom in these ways.

Before concluding this numerical analysis, we should recall one stimulus which did not affect the size of the Pernambucan slave population. There is no strong evidence for the existence of a slave-breeding industry, such as allegedly existed in the United States and which would have braked the decline. Certainly the increasing slave fertility index implies such an industry; but we do not know how much of that increase can be ascribed to regressive underenumeration of slave children (Table 24). As in Cuba, the uncertain sugar markets may have dissuaded planters from investing in slave breeding, an investment which brought returns only fifteen years later when the slave child reached his most productive working age.[23]

We can now summarize the effects of various pressures limiting and reducing the legal size of the Pernambuco slave population after 1850 (Table 28). High slave mortality, final abolition, the Free Womb Law, and the end of the African slave trade were most important. The interprovincial slave trade, followed perhaps by private manumissions and emancipations, were the next most important pressures. Least important were the Sexagenarian Law and the two emancipation funds.

All these factors were significant to the slave-owners, who thereby lost, with or without their consent, slave labor. Only certain manumissions and emancipations, on the other hand, really helped the slaves themselves, for going to work in coffee instead of sugar meant no change for the better, and possibly a change for the worst.[24] Similarly, being free in legal terms but obliged to work like a slave for one's mother's master entailed no change for the child born after 1871.

22. For examples of post-manumission qualifications, see "Libertações," *Diário de Pernambuco*, October 13, 19, November 15, 1887, April 3, 1888.

23. Alfred H. Conrad and John R. Meyer, *The Economics of Slavery and other Studies in Econometric History* (Chicago, 1964), pp. 68-69. Roland Ely, *Cuando Reinaba Su Majestad El Azúcar* (Buenos Aires, 1963), pp. 488-490. C. B. Otoni, cited in Joaquim Nabuco, *O Abolicionismo, Conferências e Discursos Abolicionistas* (São Paulo, 1949, first published 1883), pp. 89-90.

24. Bello, *Memórias de um Senhor de Engenho*, p. 45. Freyre, *The Mansions and the Shanties*, p. 131.

The total effect of pressures in the 1870's—the interprovincial slave trade, manumissions, Imperial Emancipation Fund, and natural decrease—was an annual depletion of at least 3,500 slaves. In the later 1880's, the annual rate may have risen to

TABLE 28

SUMMARY OF PERNAMBUCO'S SLAVE POPULATION

Pressures Depleting Slave Population [a]	Numbers Affected		
	Annual Average	Total [b]	Share in Grand Total (percent)
1850 Law (ends international trade,	(1,100	41,800)	- [c]
promotes interprovincial trade)	1,000	31,000	12.6
1871 Law (frees newborn of slaves,	2,266	36,807	14.9
creates imperial emancipation			
fund)	159	2,632	1.1
1883 Law (creates provincial emancipation fund)	23	135	0.1
1885 Law (free sexagenarians)	3,602	9,604	3.9
Private Manumissions with Conditions	233	8,497	3.4
Private Manumissions without Conditions	349	12,746	5.2
Deaths September 1850-September 1871	3,412	71,650	29.0
Deaths September 1871-May 1888	1,989	32,819	13.3
1888 Law (abolishes slavery)		41,122	16.7
Grand Total		247,035 [d]	100.0

[a] The slave population in 1850 was 145,000. In the average population of 126,368 between 1850 and 1873, a birth rate of 30.6 per thousand would have yielded a total of 82,488 slaves born by 1871. The maximum number of slaves available between 1850 and 1888, before considering depletions and including the ingenous born after 1871, would equal 264,295 (145,000 + 82,488 + 36,807). These calculations revise Eisenberg, "Abolishing Slavery," Table V.

[b] Totals based on averages times years and fractions of years.

[c] Since this group is hypothetical, comprising those Africans who might have been imported between 1850 and 1888, it is not included in the grand total.

[d] The difference between the grand total and the maximum available number of slaves derives from errors and omissions in reported figures and our estimates of manumissions and deaths, most likely the latter, since slaves undoubtedly died faster than the minimum death rate.

6,400 slaves as a result of the Emancipation Funds, the Sexagenarian Law, manumissions and emancipations, and natural
decrease. If all these came from sugar plantations, which numbered around 1,500 in the period, then each engenho lost about
three or four slaves yearly.

Slaveowners' Resistance to Abolition

Some Pernambucans resisted the pressures depleting their
slave labor force, but with little success. Smugglers attempted to
circumvent the ban on international slave trading, and actually
landed several hundred slaves in southern Pernambuco and Alagoas, according to the watchful British consuls. In 1855, a smug-
gler landed 200 to 240 slaves at the Barra de Sirinhaém. Two
sugar planters captured a group of these slaves; they sold some to
the provincial government, and kept others for themselves. The
British consuls complained repeatedly that the local police acted
in collusion with smugglers and planters.[25] Irritated by the incident and British charges of lax enforcement, the provincial government increased vigilance: "The entire coast is watched by
patrols ranging along it, and they do not lose sight of any ships
which seem to be navigating suspiciously. The cruiser has been
active and ready. . . .A flag-telegraph line along the coast is
being organized by the workmen under the port director, and I
believe that with eight stations to the north and ten to the south
of this capital [Recife], we will have enough means of communica-
tion for events all along the coast."[26] This watchfulness, and that
of the imperial government and the British Navy, succeeded,
and no more smuggling was reported.

25. Christopher to Palmerston, Pernambuco, February 12, 1851, *Parliamentary Papers*, 1851, HCC, v. II, *AP*, v. XXIV, Part II, pp. 498-499. *Idem.* to
idem., Pernambuco, March 12, 1851, *ibid.*, 1852-53, HCC, v. CIII, *AP*, v.
XLVII, Part II, p. 377. Burnett to Christopher, Maceió, July 10, 1851, *ibid.*,
pp. 389-390. Cowper to Clarendon, Pernambuco, November 3, 1855, and January 18, 1856, *ibid.*, 1856, HCC, v. LXII, *AP*, v. XXV, pp. 242-248. *Idem.* to
idem., Pernambuco, October 20, November 18, 1856. and January 24, 1857,
ibid., 1857, HCC, v. XLIV, *AP*, v. XX, pp. 247-264. Pereira da Costa, *Anais
Pernambucanos*, v. IX, pp. 389-390, recounts the smuggling incidents.
26. *Relatório . . . 1857 . . . Sérgio Teixeira de Macedo*, p. 22.

Stealing slaves constituted another illegal method for the planters to maintain their supply. A provincial president in the mid-nineteenth century noted: "the theft of slaves, Sirs, was so to speak a branch of trade for these merchants, demoralized men, and it reached such a point that no one could feel safe with that sort of property." The thieves often stole from towns and cities to sell to engenhos where the stolen property could not be recovered, because "in most of the provinces the agrarian interests still controlled the government, the courts, and the police." The larger engenhos also stole from the smaller engenhos, sometimes under the humanitarian guise of "sheltering fugitive slaves."[27]

The authorities tried unsuccessfully to inhibit the interprovincial slave trade. The provincial government in 1852 sharply raised the tax on each slave exported from 5$000 to 100$000, and to 200$000 in 1859. Slave exports continued, however, and in the 1870's one planter asked that the tax be doubled or tripled. But the tax rate was not raised again, and slave exports from Pernambuco continued to flow until the center-south provinces took action.[28]

The planters tried to counter abolitionist propaganda with appeals for gradualism. While the planters' SAAP had discussed the question of a labor shortage at the 1878 Recife Agricultural Congress, the accelerating pace of abolitionism in the 1880's, in particular the Ceará abolitions in 1883 and 1884, provoked new outcries. In January 1883 several prominent Escada planters, all

27. Relatório que à Assembléa Legislativa de Pernambuco apresentou na sessão ordinária de 1846 o Excellentíssimo Presidente da mesma Província Antonio Pinto Chicôrro da Gama, p. 7. Freyre, The Mansions and the Shanties, pp. 48-49, 60.

28. Relatório do Inspector da Fazenda Provincial de Pernambuco José Pedro da Silva, apresentado no 3º de fevereiro de 1852 ao Excellentíssimo Sr. Víctor de Oliveira presidente da mesma província. (Pernambuco, 1852), p. 14. Collecões de Leis e Decretos e Resoluções da Provincia de Pernambuco, Tomo XI, 1847, titulo III, p. 12, Receita Provincial, art. 1º paragraph 12, cited by Emilia Viotti da Costa, "O escravo na grande lavoura," Buarque de Holanda, História Geral da Civilização Brasileira, Tomo II, v. 3, p. 156. See also Relatório . . . 1º de março de 1853 . . . Francisco Antônio Ribeiro, p. 4. Antonio Venancio Cavalcante de Albuquerque," "Agricultura ou a questão da actualidade," Diário de Pernambuco, April 7, 1877.

SAAP members, formed a Club de Lavoura to champion gradualism and oppose "intransigent abolitionism." The Club program
urged guarantees of private property, faithful execution of the
Free Womb Law, public workhouses for freed slaves, and education of slaves' children under 21 years of age. The Club member-
ship included 88 land owners, 13 tenants, a lawyer, and a doc-
tor.[29]

At the Club's first meeting speakers warned of the "abolitionist delirium from which Ceará is suffering, where slaves have
been valued at ridiculous prices, and from whose press appear
subversive doctrines which endanger the slave owners' security,
making their relations with their slaves irreconcilable and impossible." The Club then wrote the Imperial Chamber of Deputies
to request "effective resolutions which prevent propaganda,
haste, craziness, and anarchy from triumphing over the ruins of
the Law of September 28, 1871, and the ruins of the right to justice, in whose pleasing shadow Brazilian agriculture always did
and hopes to continue to shelter itself." In four months, seventeen
Pernambuco counties held organizational meetings, and nine
clubs were founded in the province's sugar zone. The Escada
club formed an alliance with the Ipojuca club; and in mid-1884 it
approved a resolution formally joining all the clubs with the
SAAP.[30]

Having organized the anti-abolitionist planters through the
Clubes da Lavoura, the Escada planters next sponsored the Second Agricultural Congress of Recife, similar in format to the
SAAP's 1878 congress. While the 1878 congress discussed many
problems, the 1884 Recife congress focused single-mindedly on
the labor problem and the threat of immediate abolition. The
planters accused the abolitionists of inciting slave insurrection
and SAAP President Ignácio de Barros Barreto called for the for-

29. "Club da Lavoura," ibid., January 25, 1883.
30. "Acta da installação do Club da Lavoura na Escada, em 13 de março
de 1883," ibid., March 15, 1883. "Resumo da acta da 2º sessão do Club da Lavoura da Escada," ibid., April 28, 1883. "Representação da Lavoura de Pernambuco à Câmara dos Senhores Deputados," ibid., May 20, 1883. Ibid., April-July
1883, passim. SAAP Livro de Atas no. 2, August 23, 1883, June 7, 1884, July
10, 1884.

mation of private police forces to guarantee property which provincial authorities would not protect. They also opposed heavy provincial slave taxation such as had been levied in Ceará, and which they feared would raise the cost of slave-owning and create greater pressure for abolition.[31] Henrique Augusto Milet charged that abolition without indemnification destroyed capital, a resource already scarce in the northeast. Milet calculated that abolition in Ceará had cost that province 30,000 contos, on the basis of 15,000 slaves worth two contos each. Still others put the problem in political terms: Barros Barreto maintained that Ceará had in effect declared itself "a new sovereign state in South America" when it freed slaves without considering the rest of Brazil. He warned that the abolitionist campaign could split the nation, as a similar campaign had split the U.S. during the Civil War. The SAAP president compared those denying property rights to "European anarchists," and Escada mill owner Marcionilo da Silveira Lins damned abolitionists as "communists."[32]

Although the "gradualists" dominated the 1884 Agricultural Congress, some militant abolitionists were present. Manoel Gomes de Mattos, head of the Comissão Central Emancipadora, insisted that "orderly and moderate abolitionists" were not attacking individuals or inciting revolt or civil war, but only fighting against the institution of slavery. Gomes de Mattos called the Congress reactionary for not discussing any measures to achieve abolition. José Maria de Albuquerque Mello blamed the mill owners themselves for provoking slave unrest by discussing abolition in front of their slaves. He called for immediate abolition, and dismissed indemnization as "a true utopia." Isidoro Martins Júnior, who was to become a leading figure in the early republican poli-

31. "Segundo Congresso do Recife, Acta da Primeira Sessão," *Diário de Pernambuco*, July 26, 1884. This newspaper carried the main published account of the congress, June 26 through August 2, 1884; the SAAP could not afford to publish the proceedings separately. "Sociedade Auxiliadora da Agricultura," *ibid.*, February 12, 1886. Um agricultor abolicionista, "A emancipação dos escravos," *ibid.*, July 16, 1884.

32. "Sociedade Auxiliadora de Agricultura de Pernambuco," *ibid.*, July 16, 1884. "Segundo Congresso do Recife, Acta da Primeira Sessão," *ibid.*, July 26, 30, 1884.

tics in Pernambuco, supported José Maria, and suggested that the unsettling effects of abolition could be cushioned by strong work laws, repression of vagabondage, military colonies, and measures favoring immigration. José Adolpho de Oliveira Lima, a Recife merchant who later built a usina, accused the planters of arrogance for having held their congress in the ACBP meeting hall and having presumed to speak in the name of agriculture and commerce, when only a few merchants actually attended the congress. "The planters, accustomed to treat the Negro with a whip, to speak from above to the slave below, to impose their will on their domain, think that everybody should bend before their power."[33]

The planters and merchants had mixed feelings about final abolition of 1888. The SAAP insisted on indemnization for slaves freed, more rural police to control free workers, and increased treasury emissions. The ACA, which had previously supported abolition only if indemnity were granted, did not repeat this demand, but instead quickly congratulated the imperial government and proudly pointed out that the president of the reigning Council of Ministers, João Alfredo Correia de Oliveira, was a native Pernambucan.[34]

The last way in which the planters might have resisted the gradual depletion of their slave labor force was to improve the treatment of their slaves. But the available evidence, whether theoretical, impressionistic, descriptive, or statistical, does not suggest that such improvement occurred. Certainly at least three motives encouraged such an improvement after 1850. First, the end of the African slave trade meant that any increase in the slave labor force would have to result from either inter-provincial

    33. "Publicações Solicitadas," *Jornal do Recife*, July 19, 1884. "Segundo Congresso," *Diário de Pernambuco*, July 30, 1884. Um commerciante, "Congresso Agricola," *Jornal do Recife*, July 19, 1884.
    34. "Sociedade Auxiliadora da Agricultura de Pernambuco, Acta da Sessão do Conselho Administrativo havida em 4 de Novembro de 1885," *Diário de Pernambuco*, January 5, 1886. "Sociedade Auxiliadora da Agricultura," *ibid.*, July 22, 1888. "Relatório da Directoria da Associação Commercial Agricola de Pernambuco, Lido em Sessão da Assembléa Geral de 14 de abril de 1884," *ibid.*, May 21, 1884. "Associação Commercial Agricola," *ibid.*, May 17, 1888.

Plate 1. Engenho de Torre: exterior, boiler room, and horse-powered vertical mill rollers. Located in a Recife suburb. (*James Henderson*, A History of Brazil, *London, 1821.*)

Plate 2. Interior of Engenho Carauna, Jaboatao, showing horizontal mill patented by De Mornay in 1851. (*William Hadfield*, Brazil, the River Plate, and the Falkland Islands, *London, 1854.*)

Plate 3. Henrique Marques Lins, head of the wealthiest family in Escada's sugar oligarchy. *(Museu do Açúcar)*

Plate 4. Usina União e Industria, Escada. View during the first harvest, 1895-96, showing railroad cars delivering cane. *(Archive of the Usina.)*

Plate 5. Antonia Francisca da Silveira Lins, first wife of Henrique Marques Lins. *(Museu do Açúcar)*

Plate 6. The first locomotive in Pernambuco, operated by the Recife and San Francisco Railway after 1857. *(Gilberto Ferrez,* Velhas Fotografias Pernambucanas.*)*

Plate 7. Recife's inner harbor, 1862. Sailing canoes in the foreground, coasting vessels in the middle ground, ocean sailing ships in the rear. *(Bancroft Library, University of California, Berkeley.)*

Plate 8. João Felix dos Santos and daughters Luiza (l.) and Mâgina (r.). *(Museu do Açúcar)*

Plate 9. José Pereira de Araujo and
son, José, two generations of
Escada sugar oligarchs. *(Museu do Açúcar)*

Plate 10. Escada and environs, ca. 1910. Names in slanted type indicate principal engenhos
and usinas. *(Sebastião de Vasconcellos Galvão.)*

Plate 11. Slave Belisario, Engenho Cachoeirinha, Vitoria de Santo Antão; owned by the progressive planter Manoel Cavalcante de Albuquerque. He was "freed many years before abolition. He was completely trusted by the family and never left the property." (*Museu do Açúcar*)

Plate 12. Owner's residence, Engenho Matapiruma. First mill built by Henrique Marques Lins in Escada. Rear buildings were probably former slave quarters. (*Museu do Açúcar.*)

Plate 13. Mill buildings, Engenho Matapiruma. *(Museu do Açúcar)*

Plate 14. Slave laundress, Engenho Guararapes, Jaboatão.*(Museu do Açúcar)*

Plate 15. Salon, Engenho Morenos. A property in Jaboatão, owned by the Souza Leão family. (*Museu do Açúcar*)

Plate 16. Slave Mônica, nursemaid. This subject, with the child, was a common picture in nineteenth-century family photograph albums. (*Museu do Açúcar*)

trade or natural increase; the latter depended largely upon how the slave mothers and children were treated. Second, harsh treatment of any slave would obviously shorten his working life, a loss compounded by the high replacement cost. Finally, the high slave prices in the center-south could have induced northeastern slave owners interested in the inter-provincial slave trade to avoid reducing the value of their human property with mistreatment.

A more sophisticated argument leading to the same conclusion postulates that treatment depended upon market prosperity and the degree of rationalization in master-slave relationships. Both Brazilian and North American scholars have contrasted the aristocratic, paternalistic, seigneurial system on Pernambuco's plantations with the more entrepreneurial, capitalist system in the newer coffee plantations of the center-south:

The nature of relations between master and slaves varied depending upon the level of development in different areas. Relations were humanized in areas where the commercial economy underwent a crisis. . . . Relations became difficult in areas where it had become necessary to extract the maximum productive capacity from the slave. As the capitalist character of the agrarian enterprise increased, and the plantations became specialized, the traditional forms of accommodation disappeared and tensions between masters and slaves augmented.[35]

This suggestion that planters with troubled sugar markets treated their slaves better than did planters with booming coffee markets echoes the slave-owners' own conceit.[36] It also implies

35. Viotti da Costa, *Da Senzala à Colônia*, p. 460. See also Genovese, *The World the Slaveholders Made*, p. 81; and Octavio Ianni, *Raças e Classes Sociais no Brasil* (Rio de Janeiro, 1966), pp. 79-80. Another version of this argument holds that workers' welfare varied inversely with plantation size. Koster, *Travels in Brazil*, v. II, pp. 237-238, Barbosa Lima Sobrinho, *Problemas Econômicos e Sociais da Lavoura Canavieira*, pp. 225-227.

36. Koster, *Travels in Brazil*, v. II, p. 241. Bello, *Memórias de um Senhor de Engenho*, pp. 43-44. Oliveira Lima, *Memórias (Estas minhas reminiscências. . .)*, p. 119. Freyre, *The Mansions and the Shanties*, p. 131. Koster refers to slaves on rice plantations in Maranhão, where a small boom occurred in the early nineteenth century. The other writers contrast slaves in sugar with slaves in coffee.

that irrespective of the slave supply, the general decline in the Pernambuco economy after 1850 would have been sufficient to improve slave treatment.

Some historians have assumed that these considerations actually governed planter behavior.[37] But others have hypothesized that the planters increased exploitation in an attempt to maintain production levels with the reduced slave labor force.[38] Economists have devised formulas which indicate that the treatment question might depend upon simple cost-benefit calculations. The sugar planter might increase slave production in the long run by treating his slaves kindly and thereby prolonging their working lives. In the short run, however, that kindness hurt productivity—by reducing working hours, for example. The planter contemplating milder treatment thus had to compare the benefits of increased production over the long run with the costs of reduced productivity in the short run. Two criteria would determine his decision: his expectations about future income, and his costs. The first depended in part upon sugar prices and his access to markets. Since prices were falling throughout the later nineteenth century and access suffered particularly after 1870, the planter may have discounted heavily his future income stream, and chosen to maximize short-run earnings. Secondly, the interest rate, the capital cost, and the slave price or labor cost were high. Thus planters using slaves faced high carrying costs. These planters might then decide against better treatment, which would yield a smaller return in the short run and entail many years of high costs while income declined, in favor of harsh treatment, which would maximize present returns while sugar markets were still rewarding.[39] Cost-benefit analysis, fitted to data, could show whether slave-owners could have increased profits by treating their slaves better after 1850. But the problem still remains as to what actually happened.

37. Haring, *Empire in Brazil*, p. 86. Bethell, *The Abolition of the Brazilian Slave Trade*, p. 375.

38. Furtado, *The Economic Growth of Brazil*, p. 129.

39. This cost-benefits approach was suggested by David Denslow, "Slave Mortality," (mimeographed, 1969) and Nathaniel Leff, "Economic Retardation in Nineteenth-Century Brazil," (mimeographed, 1970).

The evaluation of slave treatment has challenged many students.[40] Conclusions depend upon various kinds of evidence. The most traditional sort of evidence has been impressionistic, often ambiguous reports by contemporary observers. Koster remarked that individual masters could be quite cruel, and gave examples. But he also observed that "the comforts of slaves in different situations are widely disproportionate," and while condemning the institution he refused to specify whether more slaves led "an existence of excessive toil and misery," or "a comparatively easy life." Tollenare reported less equivocally an impression of generally poor treatment.[41]

In the 1840's the comments of foreigners continued to be mainly negative. The British engineer De Mornay reported that planters worked their slaves in twelve-hour shifts seven days per week during the harvest, with no time off for meals; many slaves fell asleep on the job. De Mornay added that some planters competed to make the most sugar with the least number of slaves. British consuls in Recife reported various atrocities to their Home Office: one noted that emasculation was a "favourite" punishment for recalcitrant male slaves, while vinegar injections in the vagina castigated obstreperous females. Sadistic mill owners boiled alive both men and women. Since their government was then actively attempting to end the Brazilian slave trade, official British observers might be suspected of special pleading. Yet the French aristocrat Comte de Suzannet also told of cases of "revolting

40. For the more recent efforts by North Americans in this area, which review much of the earlier work, see Laura Foner and Eugene D. Genovese (editors), *Slavery in the New World. A Reader in Comparative History* (Englewood Cliffs, N.J., 1969); Richard Graham, "Brazilian Slavery Re-examined," *Journal of Social History* (New Brunswick, N.J.), v. 3, no. 4 (Summer 1970), pp. 431-453; Carl N. Degler, "Slavery in Brazil and the United States: An Essay in Comparative History," *American Historical Review* (Washington, D.C.), v. LXXV, no. 4 (April 1970), pp. 1,004-028; idem., *Neither Black nor White*. These writers compare slave treatment in Brazil with slave treatment elsewhere, or between different regions within Brazil. Our problem is different: did changes in treatment occur over time in the same place?

41. Koster, *Travels in Brazil*, v. II, pp. 173-174, 195, 230. When discussing punishments on plantations, Koster regretted that "though great cruelties are not *often* committed, still the mode of punishment produces much suffering, much misery, much degradation," *ibid.*, v. II, p. 223. Koster's emphasis. Tollenare, *Notas Dominicais*, p. 85.

humanity," and concluded that Brazilians lacked a moral sense.[42]

After 1850, if economic pressures led to improved treatment of slaves, reports of harshness should have disappeared. But the British consul in 1855 wrote London that "no perceptible amelioration has taken place in the condition of the slave notwithstanding the extraordinary increase in the value of this species of property." That same year, during the cholera epidemic, an official provincial committee criticized the continuing poor treatment.

The plagues kill many [slaves] but [they die] in large part due to poor treatment and little care, during the march to the fatal end. Rare are the mill's slave quarters [which are] built and placed according to hygienic principles. When they are not really badly built and located, then the slaves have not proper beds or bedclothes. Since they already do not take much care of themselves, the result is that they do not even try to avoid what can harm them. On the contrary, they desire the plagues in order to get leisure hours. There are also slave-owners who oblige them to work even though they are gravely ill. . . . If the slave-owners gave them good treatment, selecting their food and giving them sufficient and appropriate clothes; if they did not oblige the women slaves to work during the last three months of pregnancy, and had rooms where the babies would stay during infancy to receive the mothers's care or that of responsible people and be vaccinated in time, only beginning to work after 10 years of age, and only in a job which is in accordance with their strength; if during sicknesses they consulted doctors in the earliest stages, and maintained good infirmaries with the prescribed medicines; if they located the slave quarters properly, built according to the rules of hygiene, and required the slaves to clean their quarters and to bathe themselves after work, stimulating them during the winter with a rum ration: then they would see the plagues diminish considerably, and therewith mortality. This problem of the vice of eating earth would disappear, or would be considered for what it is, a symptom of need for affection. The slaves would have more vigor, and therefore their work

42. Report by De Mornay originally published in John MacGregor, *Commercial Statistics* (London, 1847-48), translated in Fernando da Cruz Gouveia, "Os De Mornay e a Indústria Açucareira em Pernambuco," *Brasil Açucareiro*, anno XXXV, v. LXX (August 1967), pp. 83-84. Cowper to Aberdeen, Pernambuco, August 4, 1843, *Parliamentary Papers*, 1844, HCC, v. XLIX, *AP*, v. XVIII, p. 366. Conde de Suzannet, *O Brasil em 1845 (Semelhanças e Differenças após um século)*, translated by Marcia de Moura Castro (Rio de Janeiro, 1957), p. 204.

would be doubled and done with more satisfaction. The number of babies born alive would be higher, and these would not die at such a high rate.[43]

A generation later, similar conditions were common. A newspaper related that "there are landowners . . . who choose the contaminated food for their slaves, and even then give each slave a ration of only one piece of jerked meat, a bowl, and a cup of poor quality meal. And the landowner does not understand that this entails a sacrifice of his capital." To cut costs, some slave-owners even refused to provide the slave with these minimal rations of food and clothing. Instead, planters traditionally had conceded the slave the right to work for himself one day per week, usually on Saturdays, Sunday, or a holiday, with the understanding that he would keep himself fed and clothed on his earnings. But "it is not believable that the slave with one day of paid work per week can subsist and acquire enough clothing to dress himself." Consequently, these slaves resorted to stealing food and clothing from other slaves and free workers.[44]

43. Cowper to Clarendon, Pernambuco, July 18, 1855, *Parliamentary Papers*, 1856, HCC, v. LXII, *AP*, v. XXV, p. 237. *Relatório da Comissão . . . Joaquim d'Aquino Fonseca. . . 10 de janeiro de 1856*, pp. 28-30. The commission recommended these humanitarian measures in the interest of preserving slave labor, not in the interest of the slaves. They suggested "instead of barbarous bodily punishment. . . other corrective means such as reducing food rations for several days, dark prisons, etc." The "vice of eating earth" was a form of suicide; it constituted a need for affection only in children. Freyre, *The Masters and the Slaves*, pp. 90, 101, 389-390, 294, 474. On the other hand, some earth also contained nutritious minerals. Josué de Castro, *Geografia da Fome (O Dilema Brasileiro: Pão ou Aço)*, 9th ed. (São Paulo, 1965, first published 1946), p. 139-140.

44. Quotes in "Breves considerações," *Diário de Pernambuco*, April 27, 1876. A precursor of this practice dates from at least the seventeenth century. The planters required the slaves to cultivate food crops on holy days and on Saturdays during the winter. This work schedule became known as the "Brazilian system." J. A. Gonsalves de Mello Neto, "Um Regimento do Feitor-Mor do Engenho em 1663," *Boletim do Instituto Joaquim Nabuco de Pesquisas Sociais*, no. 2 (1953), pp. 80-87, cited in Correia de Andrade, *A Terra e o Homem do Nordeste*, pp. 80-81. Diégues Júnior, *População e Açúcar*, pp. 69-71. Other senhores de engenho allowed the slaves to work for themselves during their off-hours—that is, from 4 to 6 in the morning and from 6 to 10 in the evening. This system was called *quinguingu*. Cruz Gouveia, "Os De Mornay e a Indústria Açucareira," p. 83.

These generally negative impressions contrast with other contemporary reports, and with most twentieth century recollections by sugar planters' descendants. Two English visitors in 1882 reported that "the poor slave . . . is better off than anyone else in the country. He is well treated in every way . . . he is a gay fellow, the only lively piece of humanity in the country, always on the broad grin and generally singing." The prominent abolitionist Joaquim Nabuco, who grew up on the Engenho Massangana in Cabo, recalled that slavery

remains in my memory as a mild yoke, the external pride of the master but also the internal pride of the slave, something like an animal's dedication that does not change because the irritation of inequality cannot disturb it. But I also fear that particular type of slavery existed only on the oldest plantations, operated for successive generations with the same humanitarian spirit, where a long legacy of fixed relations between master and slaves had included both in a type of patriarchal tribe isolated from the world.

Similarly, Júlio Bello, raised on the Engenho Queimadas in Barreiros, related that his father had given his slaves livestock and land to grow crops, and earn money: "this standard of the good, humanitarian, and generous slave-owner was the type who incontestably dominated among the old rural landowners of Pernambuco. The bad slave owners, incapable of piety and mercy, intransigent and cruel in punishing slaves, were pointed at here and there with almost general condemnation." Bello also narrated several anecdotes about senhores de engenho known for their "excessive" kindness, "who by nature never punished and thus allowed disorder and anarchy to dominate the plantation." Júlio Bello's nephew, José Maria Bello, who grew up on the Engenho Tentúgal in Barreiros, wrote that as a child he had heard stories of the "terrible cruelties of some masters: fierce beatings, tortures worthy of Chinese, such as the stocks, the crucifixions on the arms of great scales, the imprisonments in the hot stoves where the sugar was dried, castrations, etc." Nevertheless on his own plantation "the tradition was tolerance and kindness on the part of the old masters." Manoel de Oliveira Lima, a Portuguese who married into an Escada sugar family, agreed that "if in Pernambuco . . . there were bad-natured plantation owners, in compensation there also were those good-natured ones, and my father-in-law, Manuel Cavalcanti de Albuquerque as well as his brother Ambrosio Machado da Cunha Cavalcanti, were those for whom the slave was a creature with a soul."

Adelia Pinto, on the other hand, raised on plantations near Escada, heard "so much talk about how the poor slaves suffered in the hands of certain masters, that I developed an absolute horror of slavery." But Pernambuco's internationally known sociologist Gilberto Freyre, a grandson of senhores de engenho from Escada, Ipojuca, and Sirinhaém, found that "many of the big landowners soon learned that the energy of the African in their service, when abused or subjected to strain, paid less dividends than when it was well conserved." He claimed that "if there were planters who calculated the value of slaves in terms of their maximum production, killing them with hard labor, making ten do the work of thirty, the majority did not have this overwhelming lust for gain, nor this industrial concept of agriculture, and most of what they made on their cane and coffee went into the upkeep of their Negroes."[45]

These impressions, evidently inconclusive, constitute but one source of data on slave treatment. Another source is contemporary descriptions of the slaves themselves. Between 1873 and 1881, nearly one in ten fugitive slave advertisements appearing in the Diário de Pernambuco mentioned that the fugitive could be identified by "marks on the feet from hot irons," "scars on the back from whiplashes," "burns on the stomach and chest," "whip stripes on the buttocks and chest," and similar wounds.[46] The

45. Burke and Staples, Jr., Business and Pleasure in Brazil, p. 117. Joaquim Nabuco, Minha Formação (Rio de Janeiro, 1957, first published 1900), p. 188. Bello, Memórias de um Senhor de Engenho, pp. 43-44. José Maria Bello, Memórias (Rio de Janeiro, 1958), p. 16. Oliveira Lima, Memórias (Estas minhas reminiscências. . .), p. 119. Adelia Pinto, Um livro sem título (Memórias de uma Provinciana) (Rio de Janeiro, 1962), p. 175. Freyre, The Masters and the Slaves, p. 65. Idem., The Mansions and the Shanties, p. 23.

46. Our sample is from Diário de Pernambuco, January, May, and September for 1873-74, 1880-81, and March, July, and November, 1875-79. See also Gilberto Freyre, O Escravo nos anúncios de jornais brasileiros do século XIX (Recife, 1963) pp. 215-224. About three-quarters of these fugitives had es-

frequency of descriptions of brutalized fugitives at such a late date hardly indicates that masters were taking good care of those slaves. Moreover, slave revolts *en masse*, a possible index of poor slave treatment, occurred several times after 1850: in 1853 and 1867 in Pau d'Alho such revolts reached sizeable enough proportions to be mentioned in the provincial president's report. Another revolt plotted by slaves on four plantations in Recife and Cabo in 1867 was narrowly aborted.[47]

A final source of information about slave treatment is census data. If slave life expectancies increased over time, or if their mortality rates declined, these changes would support the argument that treatment improved after 1850. But incompleteness and inaccuracy in the data prevent our relying on the nineteenth century censuses. The age distribution data in the pre-1872 censuses were extremely inaccurate; the 1872 census itself was fairly complete, and can be corrected where clearly in error, but the subsequent censuses of 1890 and 1900 were deficient. The published 1890 data giving population by age did not discriminate by sex or race, and the 1900 data did not discriminate by race. These gaps prevent comparisons of the colored populations.[48]

The evidence on slave treatment does not allow concluding that planters treated their slaves better after 1850, although theoretical arguments can be made to presume as much. On balance, planters do not seem to have resisted the gradual depletion of slaves by bettering conditions, just as their resistance to pres-

caped from sugar plantations: 55 percent of the advertisements requested that the fugitive be returned to a plantation, and another 15 to 20 percent requested that the fugitive be returned to a comisario or warehouser in Recife, who presumably would either keep the slave against the planter's debts, or return him to the plantation with the next shipment of supplies.

47. *Relatório. . . 1854. . . José Bento da Cunha e Figueiredo*, p. 4. *Relatório apresentado à Assembléa Legislativa provincial em 15 de abril de 1867 pelo Exm. Sr. Conselheiro Francisco de Paula da Silveira Lobo*, p. 4.

48. Ironically, despite the absence of any specification by sex or race, the 1890 census takers carefully detailed the population by one-year age groups. I am indebted to John Knodel and Betsy Kuznesof for aid in the manipulation of these censuses.

sures depleting their slave supply did not assume large proportions.

This apparent indifference to the disappearance of the slave
labor supply derived from the fact that, planter protests notwithstanding, abolition did not create a labor crisis. Sugar production
doubled between the late 1840's and the late 1880's, despite the
fact that the slave population declined by nearly 70 percent. The
last harvest under slavery yielded a record crop, and although
production fell off 40 percent in the next two harvests, succeeding
crops returned quickly to normal levels and 1894 produced a new
record. Thus the increasing scarcity of slaves did not contribute to
the industry's stagnation. While the planters voluntarily gave up
slaves through the interprovincial slave trade, manumissions,
and emancipations, they retained control of the labor supply.
Abolition was nearly painless.

8

CONVERSION TO FREE LABOR

*T*he Pernambuco planters did not vigorously resist the gradual abolition of slavery because cheap free labor was readily available. They substituted free labor for slaves without making any material concessions to the free workers, few of whom before or after 1888 lived better than the slaves themselves. While some persons tried to improve the quality of free labor by promoting foreign immigration, none of these schemes succeeded for long. In fact the schemes were unnecessary; the Brazilian free worker met the planters' demand for labor.

Free Workers

By the middle of the nineteenth century, as we saw in the preceeding chapter, slaves had normally outnumbered free laborers on sugar plantations by ratios of over 3:1. But by 1872 free workers outnumbered slaves in all occupational categories: the free-to-slave ratio was highest in unskilled labor, 14:1, and lowest in agricultural labor, 5:1, and in domestic service, 3:1.[1]

Many of these free workers were ex-slaves who remained in the sugar zone, despite certain evidence to the contrary. Contemporary observers reported that considerable numbers of ex-slaves

1. *Relatório. . . 1854. . . José Bento da Cunha e Figueiredo*, Table. *Recenseamento da população . . . 1872*, v. XIII, pp. 1-114. Slave labor may still have predominated on sugar plantations in the center-south. See Toplin, *The Abolition of Slavery in Brazil*, p. 35.

fled their plantations, especially in the 1880's.[2] Moreover, comparison of the 1872 and 1890 censuses reveals a relative decline in the zona da mata population. In the absence of any significant immigration and presuming equal regional rates of natural increase, this change reflects a certain redistribution of population, including ex-slaves, away from the sugar zone and into the western regions of the *agreste* and the *sertão*. But colored persons probably migrated less than whites, because the density of colored persons in the sugar zone outside Recife increased slightly, while the percentage of colored persons in Recife's population remained constant—that is, the ex-slaves did not flock west or to the capital in search of greater opportunity (Table 29).

2. SAAP, *Boletim Fascículo no. 1*, pp. 30-31. Cohen to Salisbury, Pernambuco, August 14, 1889, *Parliamentary Papers*, 1890, HCC, v. LXXIV, *AP*, v. XXXIII, p. 118. For planter complaints about ex-slaves leaving the engenhos, see SAAP, *Trabalhos do Congresso Agrícola*, pp. 223, 240, 278, 315. An Olinda Law School professor believed the slaves leaving rural areas were part of a general urbanization process. "O dr. José Antonio de Figueiredo ao público, XVII," *A Província*, January 30, 1875.

TABLE 29

POPULATION DISTRIBUTION BY COLOR AND REGION

Color and Region	Census Years	
	1872	1890
Total Population Zona da Mata as percent of Total Population Pernambuco	54%	46%
Colored Zona da Mata as Percent of Colored Pernambuco	55%	52%
Colored Recife as percent of Total Recife	55%	56%
Colored Zona da Mata less Recife as percent of Total Zona da Mata less Recife	84%	87%

SOURCES: *Recenseamento da População . . . 1872*, v. XIII, p. 214. Directoria Geral de Estatística, *Sexo, raça e estado civil . . . 31 de dezembro de 1890*, pp. 94-99.

The flight from the plantations, therefore, did not mean a mass flight from sugar. At the most, it meant a redistribution of the ex-slave population within the zona da mata, with slight leakage to the west. Twentieth-century recollections by sons of senhores de engenho confirm this impression. Júlio Bello recalled that the ex-slaves "remained in the fields, respecting the white man and serving him with almost the same humility that habit and heredity had taught them." José Maria Bello added that "with abolition, the slave of Tentúgal, as those of other engenhos, abandoned the old lands of their masters, dispersing into the nearest towns and small cities, and even to Recife. Liberation for their primitive mentalities meant freedom from the plow and the field, loafing without purpose, [being] hungry and drunk on cheap rum. A little while afterwards, most of them returned, worn by misery, to the protective shadow of the Big House."[3]

The ex-slaves remained in the zona da mata because there was no place else to go. "There was already a surplus population weighing heavily upon the urban zones, and that surplus had been a serious social problem since the early nineteenth century. In the hinterland the subsistence economy had been expanded to great extremes, and the symptoms of population pressure were already being unmistakably felt in the semi-arid *agreste* and *caatinga*. These two natural barriers imposed limits on the mobility of the mass of newly freed slaves in the sugar region." Transportation costs probably prevented the ex-slaves from migrating south to the coffee areas.[4]

Free workers from the agreste and the sertão also replaced

3. Bello, *Memórias de um Senhor de Engenho*, p. 39. J. M. Bello, *Memórias*, pp. 11-12.

4. Quote from Furtado, *The Economic Growth of Brazil*, p. 151. See also Bento Dantas, "A Agro-Indústria Canavieira de Pernambuco: As Raizes Históricas dos seus Problemas, Sua Situação Atual, e Suas Perspectivas," (mimeographed, 1968), p. 15. Leff, "Desenvolvimento econômico e desigualdade regional," p. 14. Our Table 33 shows that the native population in the center-south grew faster than that in Pernambuco, a difference attributable either to prosperity-engendered higher natural increase or to internal migration. The sourse of these new Brazilians in the center-south is elusive, for the published 1890 and 1900 censuses did not include places of origin for Brazilians. Gilberto Freyre has proudly affirmed that a "biological elite" of scions from decadent oligarchical families migrated from the northeast, but these adventurous young men could hardly have comprised the bulk of the migrants. Freyre,

slaves on the plantations. When the U.S. resumed cotton exports and ended Pernambuco's brief cotton boom in the early 1870's, many free workers must have migrated east to find jobs in the sugar zone. In the later 1870's, the great droughts forced thousands of *sertanejos* to seek work on the plantations. During harvest months in the 1890's, plantations reported that as much as 45 percent of their labor forces were seasonal migrants from the sertão. Since most of these migrants maintained homes in the west, where they returned with the first rains, they did not appear in the 1890 census as a proportional increase of population in the zona da mata.[5]

The planters employed free labor in several systems, all of which kept the employee in dependency. Traditionally one of the commonest forms had been squatting. Squatters *(moradores,* also called *camumbembes)* constituted 95 percent of the free population in the southern zona da mata in Tollenare's estimate. The planter conceded the squatter a plot of poor land where he built his palm leaf or mud hut and planted a food crop such as manioc, beans, corn, or bananas. The planter protected the squatter, who often was in trouble with the legal authorities or another planter. In return for these favors, the squatter paid a share, perhaps one-tenth of his crop, and served as a guard against intruders and a pawn in the planter's recurrent feuds with the government and his neighbors. Sometimes squatters worked a specific number of days each week for the landowner, an arrangement known as *morador de condição,* which prevailed mostly in the dry mata, where plantations and slave labor forces were smaller.[6]

Ordem e Progresso, 2 vols. (Rio de Janeiro, 1959), v. 2, p. 406. Douglas H. Graham and Sergio Buarque de Hollanda Filho have written that "there was no inflow of northeastern caboclo labor to the coffee fazendas either before or after the abolition of slavery." "Migration, Regional and Urban Growth and Development in Brazil: A Selective Analysis of the Historical Record—1872-1970" (mimeographed, 1971), v. 1, p. 32.

5. Cotton was grown almost exclusively by free labor. Thomas Adamson, Jr., to William Seward, Pernambuco, November 14, 1864, in U.S. National Archives, Despatches from U.S. Consuls, v. 7. Antonio Gomes, "Agricultura—A Lavoura e o Projecto N. 7", *Diário de Pernambuco,* May 19, 1893. Correia de Andrade, *A Terra e o Homem do Nordeste,* pp. 93-94, 119-120, provides excellent summaries of rural free labor modes.

6. Tollenare, *Notas Dominicais,* pp. 95-97. Milet, "A colonisação," *Diário de Pernambuco,* May 17, 1888. Koster, *Travels in Brazil,* v. I, p. 111.

The squatter lives in fear of his landowner. He cannot be sure of awaking in the same place he lay down to sleep. He shelters himself in a miserable thatched hut built on someone else's land and conceded to him as a gift. He does not live there as long as he behaves himself and pays the price stipulated in a written rental contract, but only as long as he wants to subject himself to serve as an instrument for lust, for private vengeance, for political hatred and rivalries, and to help the landowner fight electoral battles for a cause not his own.[7]

The planters, on the other hand, justified their "violent and capricious expulsions" by citing the squatters' lack of responsibility:

This horde of men lives without rules, without morals, without respect for the rights of others, except when there is fear of the use of absolute and despotic measures, which many times the landowner finds himself in the difficult position of using. . . . At present, the landowner is at the mercy of the worker, squatter, etc., who only lives on the property until the day when he can trick the landowner. Achieving his purpose, he leaves without giving the least satisfaction to the landowner, who many times saved him from hunger and covered his bones with clothes.[8]

Wage and salary earners composed the second large group of free plantation workers. Unskilled and seasonal workers earning daily wages were the most numerous (Table 30). Long before 1888 the planters had hired wage labor, which occasionally was cheaper than slave labor (Table 31). Table 31 probably underestimates the cost differential, for the planter had difficulty finding productive tasks for all slaves in non-harvest months. Yet he had to continue paying their maintenance costs, which in the 1870's became an increasingly larger share of the cost of slave labor. Free workers, on the contrary, could be fired at a moment's notice with no further obligations. Thus most free day-workers earned

7. Paes de Andrade, *Questões Econômicas*, p. 65. For a similar and equally unhappy description, see *Falla recitada na abertura da Assembléa Legislativa Provincial de Pernambuco pelo Excellentíssimo Presidente da Provincia Conselheiro Diogo Velho Cavalcante de Albuquerque no dia 1º de março de 1871*, p. 36.

8. Cavalcante de Albuquerque, "A agricultura ou a questão," *Diário de Pernambuco*, April 5-7, 1877.

wages during the harvest months between September and March. The rest of the year, unless they were given jobs planting and cultivating cane or maintaining mill equipment, they were dropped from the plantation payroll. Salaried workers enjoying year-round employment were the least numerous on the plantations. They held the jobs requiring administrative and technical skills, and their numbers increased as the mills modernized (Table 30).[9]

During the last half of the nineteenth century, changes in salary levels for skilled and semi-skilled workers largely benefited those people; for several job descriptions, reported real salaries doubled between 1876 and 1896. Among wage earners, on the other hand, real rates generally declined despite a relatively stable population and increasing productivity (Tables 14, 22, 30, and 32.) Real wages rose slowly in the 1850's, and reached their highest levels in the 1860's and early 1870's in response to the increased demand from railroad builders. In the later 1870's, however, the droughts drove thousands from the sertão into the zona da mata, where their desperation obliged them to accept low rates.[10] Wages continued low in the 1880's owing to the increasing numbers of ex-slaves entering the labor market. In the first few years after abolition, when overnight some 40,000 ex-slaves became potential wage earners, real wages dropped to levels below those of the 1850's. The mid-1890's witnessed the highest nominal wages of the century, but the rising cost of living wiped out any real gains for wage-earners.

9. "Breves considerações sobre a agricultura no Brasil," *ibid.*, April 27, 1876. Correia de Andrade, *A Terra e o Homem do Nordeste*, p. 67. Tollenare, *Notas Dominicais*, pp. 75-76.

10. The scarcity of railroad workers was so severe that the English contractor imported 237 Belgian, Dutch, German, and Italian workers. J. A. Gonsalves de Mello, "Trabalhadores Belgas em Pernambuco (1859-1863)," *Boletim do Instituto Joaquim Nabuco de Pesquisas Sociais* (Recife), n⁰ 8 (1959), pp. 12-16. Report by Hughes, November 5, 1879, in *Parliamentary Papers*, 1880, HCC, v. LXXIII, *AP*, v. XXIV, p. 497 estimates 62,000 drought refugees. "A secca," *Diário de Pernambuco*, December 7, 1877, and a speaker at the SAAP, *Trabalhos de Congresso Agrícola*, p. 166, estimated 200,000 drought refugees.

TABLE 30

REAL WAGES AND SALARIES ON PERNAMBUCO SUGAR PLANTATIONS

(in réis)

Job Description	Number Employed	1876	Number Employed	1896	Percentage Change in Pay
1. Wage Workers (Daily)					
Cane cutters, loaders	257 (in field)	$581	15 (at mill)	$445	-23
Boilerman, fireman	6	$872	3	$612	-30
Turbine worker	3	$698	5	$834	20
Bagasse worker	4	$640	2	$501	-22
Furnaceman, fireman's helper	2	$640	3	$473	-26
2. Salaried Workers (Annually)					
Administrator	1	465$000	1	1:001$000	115
Sugar master	1	429$000	1	834$000	94
Foreman	1	334$500	1	250$000	-25
Doctor, pharmacist, nurse	1	320$000	1	1:001$000	213
Distiller	1	203$000	1	334$000	65
Bookkeeper	1	203$000	1	334$000	65

NOTE: Our data come from three different plantations; we assume comparability of job descriptions and of wage and salary scales. Where a range of payments was given, we have taken the average. All nominal values have been deflated by the 1852 price index (Table 26).

SOURCES: Presciano de Barros Accioli Lins, "Despezas de algumas experiencias feitas no engenho Tinoco sobre a manipulação da canna, seu assucar e aguardente," *Jornal do Recife*, May 5, 1876. José Bezerra de Barros Cavalcante, "Demonstração do resultado da canna, segundo o uso geral nos engenhos desta provincia," *ibid.*, July 28, 1876. "Relatório que ao Excmº Sr. Conselheiro Governador do Estado apresenta o Engenheiro Adolpho Barbalho Uchoa Cav.te Acerca dos trabalhos da Escola Industrial Frei Caneca no 1º semestre de 1896," *Diário de Pernambuco*, September 2, 16, 1896.

TABLE 31

COMPARATIVE MONTHLY LABOR COSTS IN PERNAMBUCO

(in réis)

I Interest, 15% per year [a]				
D Depreciation, 7% per year [b]			*W Minimum daily wage*	
M Maintenance (food and clothing) [c]			*T Total wage cost,*	
T Total monthly cost [d]			*25-day working month*	

Year		Slave		Free
1817	I	1$108		
	D	$846		
	M	1$083	W	$160 [e]
	T	3$137	T	4$000
1857	I	15$000		
	D	6$996		
	M	7$914	W	$978
	T	29$910	T	24$450
1862	I	10$838		
	D	5$055		
	M	5$112	W	1$043
	T	21$005	T	26$075
1874	I	5$700		
	D	2$659		
	M	5$952	W	1$000
	T	14$311	T	25$000
1880	I	8$538		
	D	3$982		
	M	6$654	W	$640
	T	19$174	T	16$000
1887	I	3$538		
	D	1$650		
	M	5$814	W	$530 [f]
	T	11$002	T	13$250

[a] Interest rates after 1850 sometimes ran higher than the 1.25 percent monthly used by Tollenare. Even at 2 percent monthly, however, the comparisons would not change (see Chapter Four).

[b] We assume the slave-owner bought the slave at the beginning of his working life, which may have averaged 15 years. This assumption is probably generous: other data suggest that the slave's average life span was no more than 27 years. See W., "Seguro contra a mortalidade dos escravos," *Diário de Pernambuco,* November 24,

1856, for an estimate of a 20-year average slave life; and Arriaga, *New Life Tables*, p. 42, for a calculation that in Brazil in general the average life expectancy at birth in 1872 was 27.4 years. A 15-year working life may have been common in coffee. Viotti da Costa, *Da Senzala à Colônia*, p. 256. Even if we assume a 30-year working life, however, which would cut given depreciation costs in half, free labor would still be cheaper than slave labor in the years noted.

[c] Clothing costs were assumed equal to one-sixth of food costs, as estimated by Tollenare, *Notas Dominicais* p. 75. We also assume retail unit food prices in the sugar zone doubled Recife bulk prices used in Table 26.

[d] Compare Roberto Simonsen, "As consequencias econômicas da abolição," *Jornal do Comércio* (São Paulo), May 8, 1938, reprinted in *Revista do arquivo municipal* (São Paulo), ano iv, v. XLVII (May 1938), p. 261, where he estimates that when slaves cost 900$000 and daily wages were $400, slave labor cost at least 50 percent more than free labor in the north. Compare also Viotti da Costa, *Da Senzala à Colônia*, p. 198, who found in Bahia in 1879 monthly slave maintenance cost 12$500 and in western São Paulo in the mid-1880's monthly slave maintenance cost 20$000.

[e] We assumed the 1817 wage equaled the wages in 1802 and 1829 listed in our Table 32.

[f] We averaged the 1886 and 1888 wages to derive the 1887 wage, Table 32.

SOURCES: Tollenare, *Notas Dominicais*, pp. 74-75. Our Tables 25, 26, and 32.

The rural wage earners suffered a decreasing standard of living in part because they did not defend themselves by organizing unions. The first decade of the Republic saw the first unions and strikes in Pernambuco's history. The first Pernambuco labor organization, Centro Operário, announced its formation in July 1890, and eight months later a group of sugar refinery workers in Recife declared the first strike to demand 50 percent pay increases. At no time during the period, however, did plantation workers unite to take similar action. Presumably their low level of class consciousness and the landowners' tight political control precluded such labor organization. Instead, plantation workers were used as scabs: three Escada usina owners offered to bring their workers into Recife to help break a 1906 strike.[11]

The third mode of free employment on plantations, numeri-

11. "Liga Operaria Pernambucana," *Diário de Pernambuco*, July 26, 1890. "Greve dos batidores de assucar," *ibid.*, March 5, 1891. In subsequent years railroad workers, cigarette factory girls, and sugar refinery workers again walked out to demand pay increases. Pires Ferreira, *Almanach. . . 1903, 1906, 1908, 1909*, see the "Cronologia" section in each. "Greve," *A Provincia*, November 17, 1906. This strike involved sugar warehouse workers, stevedores, bakery workers, trolley drivers, and butchers; the government used the police to smash it, to the applause of the commercial community. *Relatório de Direcção da Associação Commercial de Pernambuco, apresentado à Assembléa Geral da mesma em 1907*, pp. 46-58.

TABLE 32

UNSKILLED RURAL LABOR

MINIMUM DAILY WAGES FOR PERNAMBUCO

(in réis)

Year	Nominal	Real (1852=100)	Year	Nominal	Real (1852=100)
1802	$160	n.a.	1884	$800	$415
1829	$160	n.a.	1886	$500	$319
1842	$255	n.a.	1888	$560	$418
1855	$580	$330	1889	$600	$255
1856	$652	$295	1890	$500	$240
1857	$978	$459	1895	1$200	$283
1859	1$076	$432	1896	1$200	$334
1862	1$043	$756	1897	1$500	$291
1874	1$000	$625	1900	1$200	$396
1876	1$000	$581	1902	$800	$333
1880	$640	$358	1910	1$030	n.a.
1882	$800	$345			

SOURCES: *1802:* Pereira da Costa, *Anais Pernambucanos,* v. VII, p. 104. *1829: Ibid.,* v. IX, p. 313. *1842:* G. T. Snow to Daniel Webster, Pernambuco, September 1, 1843, in U.S. House of Representatives, *Executive Documents,* Second Session of 28th Congress, v. 3, D. 73, p. 247. In calculating the wage we used the average exchange rate, not prevailing par. Onody, *A Inflação Brasileira,* pp. 22-23.

1855, 1856, 1857, 1862: Report by Hunt, Pernambuco, August 18, 1864, in *Parliamentary Papers,* 1857, HCC, v. LIII, *AP,* v. XXIV, pp. 366-267. For 1856, see also Bellamy to Sherburne, Pernambuco, December 2, 1856, *ibid.,* 1857, HCC, v. XLIV, *AP,* v. XX, p. 252.

1874: Informações sôbre o Estado de Lavoura, p. 160. *1876:* Milet, *Os Quebra Kilos,* pp. 4-5. *Idem., A Lavoura da Canna de Assucar,* pp. 104-112. *1880:* "Banco Agrícola," *O Brazil Agrícola,* Anno II, November 15, 1880, p. 38. *1882:* "Bancos de Crédito Real," *Diário de Pernambuco,* August 12, 1882. *1884:* Um agricultor, "O abolicionismo e a lavoura," *ibid.,* April 6, 1884. *1886:* SAAP, Livro de Atas no. 2, February 10, 1886.

1888: Milet, "A Colonisação," *Diário de Pernambuco,* May 17, 1888. *1889:* "Representação dirigida ao governo da provincia pela Sociedade Auxiliadora da Agricultura," *ibid.,* August 13, 1889. *1890:* "Nucleo Suassuna," *ibid.,* February 7, 1890. *1895:* "Pela Lavoura," *Jornal do Recife,* July 30, 1895. *1896:* "Escola Industrial Frei Caneca," *Diário de Pernambuco,* September 16, 1896.

1897: Report on the Trade and Commerce of the Consular District of Pernambuco for the year 1897, *Parliamentary Papers,* 1898, HCC, v. XCIV, *AP,* v. 43, p. 8. *1900:* Report on the Trade and Commerce of the Consular District of Pernambuco for the years 1899-1900, in *ibid.,* 1901, HCC, v. LXXXI, *AP,* v. 45, p. 12. *1902:* Um Agricultor, "O Preço do Assucar," *Diário de Pernambuco,* January 11, 1902. *1910:* Directoria Geral de Estatistica, *Industria Assucareira. Usinas e Engenhos Centraes,* pp. 3-6.

TABLE 33

POPULATION BY ORIGIN

Area	Year	Brazilians	Foreigners [a]	Total Population	Foreigners/ Total Population (percent)
Pernambuco					
	1872	828,089	13,444	841,533	1.6
	1890	1,027,534	2,690	1,030,224	0.3
	1900	1,167,328	10,822	1,178,150	0.9
Rio de Janeiro [b]					
	1872	910,394	184,182	1,094,576	16.8
	1890	1,259,043	140,492	1,399,535	10.0
	1900 [c]	1,469,257	268,221	1,737,478	15.4
São Paulo					
	1872	807,732	29,622	837,354	3.5
	1890	1,309,723	75,030	1,384,753	5.4
	1900	1,753,092	529,187	2,282,279	23.2

[a] Includes Africans.

[b] Includes Município Neutro, later known as the Distrito Federal.

[c] Data for the Distrito Federal are for 1906.

SOURCE: Oliveira Vianna, "Resumo histórico dos inquéritos censitários," pp. 429-432, 462-463, 478-480.

cally the least important but materially the most comfortable, was sharecropping. The sharecropper (lavrador) received a plot of land to grow sugar cane; sometimes the landowner supplied cane cuttings to seed the first harvest, and permitted the sharecropper to grow a food supply for his family and slaves. The sharecropper had to plant, cultivate, cut, and transport the cane to the mill at times designated by the mill owner. He paid one-half his cane or the sugar made therefrom to the senhor de engenho, as well as all the molasses, rum, and waste products.[12]

Sharecropping had existed since the sixteenth century, and in the nineteenth-century sharecroppers were a small but growing rural middle class. Tollenare counted between two and three

12. SAAP, Trabalhos do Congresso Agrícola, pp. 323-325.

sharecroppers per plantation in the early nineteenth century; they were usually Brazilian whites, and each owned six or seven slaves. By 1842, sharecroppers owned 30 percent of the slaves on 382 plantations surveyed; a decade later they produced cane for 42 percent of the sugar in Jaboatão; and in 1878 reportedly one-half the sugar shipped outside the province came from cane grown by sharecroppers.[13] This increasing reliance on share-cropping prefigured the division of labor reembodied in the central mill and the usina. Most likely the falling sugar prices motivated the planters to minimize risks by delegating cane-growing to the sharecropper and thereby tying cane costs to the sugar price.

The planters encouraged sharecropping. "For the man unfavored by fortune there is no lack of lands for him to work at present, either through a ridiculous and insignificant rental or through a very small profit which they give to the landowners who offer all their land for them to cultivate as they see fit . . . even houses, with only the desire to see their properties inhabited and cultivated." Another mill owner pointed to the example of a sharecropper who began with two slaves, and ended his life with a title of nobility, an estate worth 1,000 contos, and nine children who acquired "immense fortunes."[14] Ten years before final abolition, Escada planter João Manoel Pontual expected sharecropping to be "the only form of future free labor in sugar production."[15]

The sharecropper himself was less enthusiastic about the system. While he paid a rent in sugar cane, he usually received no contract specifying the terms of his tenure. The mill owner

13. Tollenare, *Notas Dominicais*, p. 93. Figueira de Mello, *Ensaio sôbre a Statística*, p. 263. "Uma Estatistica," *Diário de Pernambuco*, January 4, 1858. SAAP, *Trabalhos do Congresso Agrícola*, p. 324. The speaker said one-half of the province's sugar "exports," but in the nineteenth-century usage "exports" included sugar shipped to the southern provinces. See this usage, for example, in the statistical tables published in the ACBP's annual *Relatórios*.

14. Cavalcante de Albuquerque, "A agricultura ou a questão," *Diário de Pernambuco*, April 5, 1877. Ceresiades, "A agricultura em Pernambuco, IV," *ibid.*, June 22, 1878. Such success stories were not common: perhaps one in 1,000 sharecroppers became a mill owner. L. B., "Banco Agrícola," *O Brasil Agrícola*, Anno II, November 15, 1880, p. 38.

15. SAAP, *Trabalhos do Congresso Agrícola*, p. 220.

could oblige him to accept loans at usurious rates and false weights on his cane. If the sharecropper objected, the mill owner, with several sources of sugar cane, could refuse to mill the recalcitrant's crop, which had to be ground within 48 hours after cutting to extract the most juice. The sharecropper was not allowed, and probably was not able because of transport costs, to seek an alternate buyer, nor could he afford to lose his only cash crop. If he persisted in his objections the senhor de engenho could evict him and easily find another landless individual to cut and deliver the cane. As a result of his insecurity, the sharecropper invested his surplus cash in movable assets such as slaves and livestock, which he could take with him as he "led an almost nomadic life, roaming from plantation to plantation."[16]

Several planters recommended improving the sharecropping system, as much to stabilize labor and to guarantee cane supply as to better the sharecroppers' lot. Ignácio de Barros Barreto saw "in sharecropping an effective corrective for the lack of slaves," but urged "offering certain advantages to our hired hands and sharecroppers, either guaranteeing them a better quality product from our mills, or facilitating the transport of cane to our mills or factories, and preparing the cane-fields for them when that work is beyond their capacity." Francisco do Rego Barros de Lacerda claimed to supply his sharecroppers with an elementary school, clothes, books and materials, houses, and interest on money banked with him, as well as manure, seeds, and tools. Other planters recommended reducing the mill owner's share, paying cash, selling the sharecropper his land, and writing formal contracts.[17]

16. Tollenare, *Notas Dominicais*, pp. 93, 96-97. SAAP, *Trabalhos do Congresso Agrícola*, pp. 323-324.

17. Sociedade Auxiliadora da Agricultura de Pernambuco, *Acta da Sessão da Assembléa Geral de 23 de abril de 1877 é Relatório Annual do Gerente o Senhor Ignácio de Barros Barreto* (Recife, 1877), p. 19. "Industria agricola assucareira em Pernambuco," *Diário de Pernambuco*, July 22, 1881. Tollenare, *Notas Dominicais*, p. 94. M. P., "Colonisação," *Diário de Pernambuco*, December 20, 1871. "Elemento servil," *ibid.*, March 14, 1882. "Preços do assucar e futuro de nossa industria assucareira," *ibid.*, December 10, 1886. Milet, "A colo-

Despite the examples of some and the recommendations of others, the sharecropping system endured with little change. As a result of the transition from engenho to usina, the smaller share-croppers who supplied usinas became known as *parceiros*, and they paid up to one-half their cane as rent. The larger sharecrop-pers, known as *rendeiros*, paid lower percentages which decreased as their production increased, but they were also obliged to provide a certain minimum quantity of cane and were financially responsible if they failed to do so. In one sense, mod-ernization broadened the sharecropping system to include the senhores de engenho now known as fornecedores, who supplied the central mill or usina with cane. These planters could become as dependent upon the usina owner, who financed their crop and bought their cane, as were their own sharecroppers upon them. During the harvests of 1902-07, one of the larger Pernambuco usinas, União e Indústria in Escada, bought cane from an aver-age of 18 fornecedores and 14 rendeiros working a total of 18 engenhos.[18]

Forced Labor

Although the depressed wage level indicated a relative abundance of labor, the planters complained often about the qual-ity of free workers. Accustomed to docile slave labor, the planters resented the workers' laziness and inconstancy, their refusal to work steadily for long periods. These complaints led some to call for laws repressing vagabondage and obliging men to work.

The inconstancy of free rural labor was notorious; ex-slaves and muleteers in particular declined steady employment. One

nisação," *ibid.*, May 17, 1888. "O dr. José Antonio de Figueiredo ao publico, XXI," *A Provincia*, February 11, 1875. SAAP, *Trabalhos do Congresso Agrí-cola*, pp. 315, 324-325, 380, 440-441. In 1903 the União Agrícola de Jaboatão drafted an elaborate set of regulations for sharecropping, which illustrate the continued dependency. There were 26 articles of sharecroppers' obligations, but only 7 articles of rights; only the sharecropper could be punished for failure to conform to the regulations, and his appeal would be heard by other landowners. Peres and Peres, *A Industria Assucareira em Pernambuco*, pp. 247-251.

18. Dé Carli, *O Processo Histórico da Usina*, pp. 20-21. SAAP, Livro de Atas no. 2, March 25, 1876. Archive of the Usina União e Indústria, Livros Diários for 1902-07.

planter asked "how can one work well with free labor when they are not dependable, they do not accept work contracts, and they cheat and move from one sugar mill to the next, getting drunk and stealing cane and manioc?"[19] Others deplored that "the workers in general do not have work habits, and they think that to be *free* is to have the liberty not to work, although the idleness to which they deliver themselves is highly prejudicial." In the towns and cities "the emancipated were refusing to work and preferring the vagrant life," or working only a few days each week "if they can earn sufficient to live for the remainder of the week."[20]

The muleteers, disemployed in large numbers by the advent of the railroad, especially aggravated the planters. "The muleteers, living like nomads, without the precious habits of work, without an attachment to the soil that would create a love for the home and all the emotions of the family, besides being so many work hands robbed from agriculture, constitute a seminary from which have come almost all the famous assassins and horse-thieves." The muleteers lived in what modern economists would call disguised unemployment; "whoever has traveled in the interior of our provinces must have met with these innumerable bands of muleteers, who clog the roads, and who look very like Bedouin caravans. Why this? Why are so many men employed in this transport industry, itself very unprofitable, and why . . . is

19. A. C. "Elemento Servil," *Diário de Pernambuco*, March 14, 1882. Koster had noted the custom of cane-stealing for personal consumption by free workers. *Travels in Brazil*, v. II, p. 117. In 1882 a group of planters complained that gangs of eight to ten men were regularly raiding their canefields by day and night. "Appello feito ao Dr. Chefe de policia pelos agricultores de Jaboatão," *O Brasil Agricola*, Anno IV, August 31, 1882, p. 186.

20. "De que precisa a industria, V," *O Industrial*, v. I, no. 9 (September 15, 1883), p. 99. "Parecer da secção da cultura lido pelo respectivo relator o Dr. Paulo de Amorim Salgado na secção de Assemblea Geral no dia 28 de Agosto último," Sociedade Auxiliadora da Agricultura de Pernambuco, *Boletim*, no. 3 (September 1882), pp. 27-28. Cohen to Salisbury, *Parliamentary Papers*, 1890, HCC, v. LXXIV, *AP*, v. XXXIII, p. 118. Consul Howard, "Report on the Trade and Commerce of the Consular District of Pernambuco for the Years 1899-1900," *ibid.*, 1901, HCC, v. LXXXI, *AP*, v. 45, p. 12. That free workers could survive without working a full week does not mean that wages were generous; more likely these individuals were young bachelors without dependents who supplemented their cash income with stolen produce, fish, game, and wild fruit.

their moral development so retarded?"[21]

The planters found the vagabonds so objectionable that they proposed sanctioning forced labor. As early as 1851, many people feared an extension of the definition of slave to include all colored men, and when an imperial decree ordered a register of births and deaths by color, armed revolts broke out in December 1851 and January 1852 in twelve sugar parishes. The rebels, also known as "wasps," feared that the new register would help the authorities force free colored men to work, and they gathered 1,000 men in Pau d'Alho. The provincial government dispatched two infantry battalions, but the rebels upon receiving reassurances disbanded without bloodshed.[22]

When wages climbed in the early 1870's, the idea of forced free labor again attracted considerable attention. To put the muleteers to work on the plantations, one senhor de engenho wrote that "it was only necessary that the government take steps with special laws to repress vagabondage." In 1874 an agitated group met in Escada to hear a paper "which had as its purpose to require free people to work without recompense." Monks dispersed the crowd only after convincing them that the alleged document did not exist. Even in the later 1870's, when drought refugees could be hired at half the normal wage, planters were agreeing that "a law which makes work obligatory will be very necessary," and asking "that a severe police regime be imposed, to which all individuals without trade or craft should be subjected." "Let us oblige the lazy to work," exhorted one orator at the

21. Quotes in Paes de Andrade, Questões Econômicas, p. 55; and Tavares de Mello in the Assemblea Legislativa Provincial, May 18, 1866, quoted in O Brasil Agrícola, Ano IV, no. 23 (August 25, 1866) p. 361. This magazine appeared in two series, the first in the 1860's and the second beginning in 1879. Joan Robinson first used the term "disguised unemployment." Alfredo Navarrete, Jr., and Ifigenia M. de Navarrete, "Underemployment in Underdeveloped Economies," in A. N. Agarwala and S. P. Singh (editors), The Economics of Underdevelopment (New York, 1963), p. 342. See also Herculano Cavalcanti de Sá e Albuquerque, "Elemento Servil," Diário de Pernambuco, September 22, 1871.

22. Relatório. . . Víctor de Oliveira. . . 9 de março de 1852, pp. 3-4. Mello, Pau d'Alho, pp. 16-22.

1878 Recife Agricultural Congress, because "the agglomeration of idle men in the large population centers is an imminent danger, a postponed and brutal revolution." Other speakers pointed out that forced free labor could permit the planters legally to keep the slaves' free children on the plantation. In the aftermath of final abolition the head of the Banco de Crédito Real wrote a pamphlet calling for state asylums and poorhouses, and strict penalties for vagabonds and beggars who did not enter such institutions. His proposals were seconded by the state Chief of Police.[23]

But in the era when the politicians finally abolished slavery, it was naive if not presumptuous to expect those same officials to countenance new forms of coercive labor. Moreover, even if the recommendations had been followed, they probably would not

23. Quotes in Cavalcanti de Sá e Albuquerque, "Elemento servil," *Diário de Pernambuco*, September 22, 1871. "Cidade da Escada," *ibid.*, January 18, 1874. SAAP, *Trabalhos do Congresso Agrícola*, pp. 136, 291, 450, 243. Only two planters opposed proposals to coerce labor at the 1878 Recife Agricultural Congress. *Ibid.*, pp. 149, 379. Gomes, "Agricultura—A Lavoura e o Projeto N. 7," *Diário de Pernambuco*, April 23, 1893. Paulo de Amorim Salgado, "Sociedade Auxiliadora da Agricultura de Pernambuco, Parecer do Presidente da Secção 8⁰, relativo aos meios de se debellar a crise da Lavoura," *ibid.*, September 1, 1895. *Idem.*, "A Crise Agrícola e Arbitros para sua solução," *ibid.*, May 2, 1897. Um agricultor, "A Crise Agrícola e Arbitros para sua solução," *ibid.*, May 8-9, 1897. Santos Dias Filho, "Propaganda—A Lavoura. Policia Rural," *O Agricultor Pratico*, Anno 2, no. 1 (June 1, 1904), p. 95. "Chronica parlamentar," *Jornal do Recife*, April 7, 1893. A writer in the rival *Diário de Pernambuco* feared certain measures such as contracts and workbooks would provoke "the flight on an elevated scale of the workers who still sustain, however inadequately, sugar cane agriculture." Gomes, "Agricultura—A Lavoura e o Projeto N. 7," *Diário de Pernambuco*, April 23, 1893. A distinguished planter felt any law would be futile because "the law for the rural laborer is the custom on the plantation where he happens to work." Davino Pontual, "A Agricultura de Pernambuco, I. Organisação de Trabalho," *O Agricultor Pratico*, Anno 2, no. 4 (July 15, 1903), p. 26. João Fernandes Lopes, *Colônias Industriaes Destinadas à disciplina, correcção e educação dos vagabundos regenerados pela hospitalidade e trabalho ou Exemplos fecundos das medidas preventivas contra a mendicidade e vagabundagem empregada na França, Suissa, Allemanha, Hollanda, Inglaterra e Estados Unidos por meio de regulamentos até 1889*, (Pernambuco, 1890), pp. 96-100. "Colonia Agricola Disciplinar," *Jornal do Recife*, May 3, 1890. Similar fruitless appeals for forced labor were also made in the center-south. Stein, *Vassouras*, pp. 263-264.

have produced the expected results. Vagabondage resulted less from low moral character or inadequate policing than from a lack of incentive. The depressed wage level, the scarcity of land, the planters' habits of paying in kind or in part and of charging inflated prices for staples at the plantation store—all these factors discouraged the rural worker. When a nationally respected sugar chemist warned that existing wages "obviously cannot be any incentive to the worker," he was ignored. Instead, usina owners wondered if they might not solve their market crisis by cutting wages and salaries, and shouted down a speaker at the 1905 Recife Sugar Conference who suggested replacing salaries with profit-sharing.[24]

Foreign Immigration

While planters were considering how best to re-establish forced labor, other individuals, mostly Recife merchants, were sporadically promoting foreign immigration to improve labor quality, but with equally poor results.[25] This experience contrasted sharply with the thousands of immigrants who settled in the center-south provinces in the same period. The most numerous attempts to attract immigrants to Pernambuco coincided with the gradual abolition of slavery. In 1857 a group of prominent Portu-

24. L   "A Lavoura," *Diário de Pernambuco*, August 7, 1897. Casti   d that the plantation work schedule, which required certa   ew days each week, might also contribute to vagabondage. "M   ola," *ibid.*, May 28, 1901. José Rufino, "Concessões de Usinas," *Jor.   ife*, August 6, 1895. Um vosso assignante, "Illm. Srs. Redactores do Diário de Pernambuco," *Diário de Pernambuco*, October 13, 1896. "Reunião de Agricultores em Palmares," *ibid.*, October 15, 1896. Leonardo Cavte de Albuquerque, "Movimento Agricola," *ibid.*, June 16, 1900. "A grande reunião dos agricultores," *ibid.*, December 13, 1905. *Trabalhos da Conferência Assucareira*, Part I, pp. 86-87.

25. For early nineteenth century attempts to start European immigrant colonies in Pernambuco, see Peter L. Eisenberg, "Falta de Imigrantes: Um Aspecto do Atraso Nordestino," *Revista de História* (São Paulo) (January-March 1973). This article was originally presented as a paper to the 24th annual meetings of the Sociedade Brasileira para o Progresso da Ciência, São Paulo, July 1972. I am grateful for comments by conference participants. See also *idem.*, and Michael M. Hall, "Labor Supply and Immigration in Brazil: A Comparison of Pernambuco and São Paulo," paper presented to the 4th annual meetings of the Latin American Studies Association, Madison, Wisconsin, May 1973.

guese merchants in Recife formed the Associação para Colonização de Pernambuco, Paraíba e Alagoas. The association proposed
to settle "industrious, well-behaved farming immigrants" on "the
unoccupied government lands or others in the public and private
domain." It would receive government reimbursement for subsidizing passage and keep, rentals and sales of association lands,
and interest on loans to colonists and advances to planters importing their own immigrants. The imperial government approved
the group's statutes and raised 500 contos capital, but in 1858 the
association disbanded. The imperial government in 1857 also
ordered a survey for an agricultural colony near the Alagoas border, to be operated by Belgian Trappist monks, but four years
later the project was suspended for lack of agreement with the
Trappist Superior.[26]

In 1864, a Polish aristocrat, the Count Anton Ladislaw
Jasiensky, appeared in Pernambuco to promote the immigration
of Poles. The provincial assembly obligingly created a lottery to
pay the expenses of the Associação Promotora da Colonisação Pola-
ca no Brasil. But the ACBP soon complained that "the Polish colonization with which they tried to dazzle us was no more than an
authentic hoax." At the conclusion of the Civil War in the U.S.,
four families from the defeated Confederate States actually
settled between Palmares and Garanhuns, where they began
planting cotton. The provincial government contracted an access
road to the colony, and the provincial president in 1866 reported
the little group "satisfied" and "pleased," but they never harvest-
ed a second crop.[27]

26. "Estatutos da Associação de Colonisação em Pernambuco, Parahiba e
Alagoas," Diário de Pernambuco, July 29, 1857. "Directoria da Associação para
Colonisação de Pernambuco, Parahiba e Alagoas," ibid., October 27, 1857. Pereira da Costa, Anais Pernambucanos, v. IX, p. 337-338. This source did not specify the "difficulties" which caused the association's dissolution, and the directorate's Relatório was unavailable.

27. Relatório. . . ACBP. . . 26 de novembro de 1867, p. 7. Pereira da
Costa, Anais Pernambucanos, v. IX, pp. 338-339. Relatório que o Excellentíssimo Senhor 1º Vice-Presidente Dr. Manoel Clementino Carneiro da Cunha
apresentou ao Excellentíssimo Senhor Conselheiro Dr. Francisco de Paula Silveira Lobo por occasião de entregar-lhe em novembro de 1866, a administração
da província de Pernambuco, p. 24. Milet, "A Colonisação," Diário de Pernambuco, May 17, 1888. The U.S. Consul could not persuade the Americans to stay.
Thomas Adamson, Jr., to William H. Seward, Pernambuco, August 29, 1866, in

The 1871 Free Womb Law reawakened interest in immigration. The provincial president called a public meeting to create the Sociedade Auxiliadora da Immigração e Colonização para a Provincia de Pernambuco. Leading politicians, merchants, and planters headed the society, which proposed to acquire and transfer lands to colonists, lobby with the imperial government, and maintain an immigrant shelter near Recife. The Society quickly raised 73 contos in capital, but it never did anything else. The ever-enterprising Bento José da Costa Júnior contracted the imperial government in late 1871 to introduce 15,000 colonists at so much per head over five years in the province north of Alagoas; but although his contract was renewed in 1874 he failed to produce. Joaquim Caetano Pinto Júnior signed a similar contract in 1875; his first batch of 116 immigrants soon left "saying that the salary was very little, and that since they were not farmers, the lots offered were not suitable."[28] Joaquim Lopes Machado aided 295 French immigrants sheltered at the Recife Navy Arsenal in 1875, but "the immigrants encountered such difficult circumstances that few settled, most preferring to continue on to Pará at

---

U.S. National Archives, Despatches from U.S. Consuls, v. 8. Other schemes fell through for lack of official interest. *Relatórios com que o Excellentíssimo Senhor Barão de Villa Bella passou a administração desta provincia ao Excm. Sr. Vice-President Dr. Quintino José Miranda em 23 de Julho de 1868, êste ao Excm. Sr. Vice-Presidente Desembargador Francisco de Assis Pereira Costa em 28 do mesmo mez e anno, e o último ao Excellentíssimo Senhor Presidente Conde de Baependy em 23 de agosto seguinte*, pp. 14-15. *Annaes da Assembléa Provincial de Pernambuco, Quinto Anno, sessão de 1871*, pp. 126-127.

28. "Reunião," *Diário de Pernambuco*, December 21, 1871. "Sociedade Auxiliadora da Immigração e Colonisação estrangeira e nacional para a provincia de Pernambuco," *ibid.*, January 8, 1872. *Falla com que o Exm. Presidente da Provincia João José de Oliveira Junqueira abriu a Assembléa Legislativa Provincial de Pernambuco no dia 1º de março de 1872*, p. 37. "Decreto 114," *Diário de Pernambuco*, November 29, 1871. *Falla com que o Exmo. Senhor Desembargador Henrique Pereira de Lucena abrio a Assembléa Legislativa Provincial de Pernambuco em 1º de Março de 1875*, pp. 146-147. Bento José da Costa Júnior apparently made a habit of failing to fulfill government contracts. In 1874, he contracted to build a railroad from Recife to Caruarú, but did not build it. *Falla. . . Henrique Pereira de Lucena . . . 1º de março de 1874*, p. 63. Pinto, *História de Uma Estrada-de-Ferro*, pp. 103-104. "Immigrantes," *Diário de Pernambuco*, February 26, 1875.

the government's expense, or to return to Europe at the expense of subscriptions raised with their consuls' aid." Of another 179 French immigrants also sheltered in the Navy Arsenal in 1875 only a "few were hired by agriculture, and not a single one wanted the lots surveyed."[29]

The repeated failures discouraged any further immigration schemes until the late 1880's. When the imminence of abolition raised the spectre of a labor shortage in 1888, the imperial government appointed José Osório de Cerqueira as Inspector-General of Lands and Colonization in Pernambuco, offered to pay immigrants' transatlantic passage, and opened a credit of 50 contos to aid immigrant colonies in Pernambuco. José Osório published a flattering description of the province in Portuguese, French, and Italian in an edition of 18,000 copies. The provincial president appointed a committee which formed the Sociedade Promotora da Colonização e Imigração led by Recife merchants. The following year the imperial government increased its subsidy to 120 contos, and provincial authorities bought the Engenho Suassuna in Jaboatão and the Jaqueira farm outside Recife, and printed 20,000 copies of a map of Pernambuco; Henrique Marques de Holanda was authorized to place ten European families on his Usina Mameluco. The excitement over immigration reached its peak when the immigration Inspector-General announced that 100,000 colonists had been contracted for the provinces between Bahia and Pará, and the press reported contracts for 775,000 Europeans.[30]

29. "Emigrantes," ibid., April 13, 18, 1875. Falla. . . João Pedro Carvalho de Moraes. . . de março de 1876, pp. 81. The French immigrants had originally landed in the Rio de la Plata area, probably Buenos Aires. But "the lack of security and peace" there, due to the political fighting between Bartolomé Mitre and Nicolás Avellaneda after the latter won the 1874 election, had persuaded them to try Pernambuco. The unhappy French could not have found better prospects in Pará. The capital, Belém, was a small riverside town of 15,000 in the mid-nineteenth century. The population began increasing in the later 1870's with the arrival of drought refugees from Ceará, but the economy only started to boom with natural rubber exports in the late 1880's. Pierre Denis, Brazil, translated by Bernard Miall (London, 1911), pp. 358-359. Relatório. . . Henrique Pereira de Lucena. . . 19 de maio de 1875. Lucena blamed export price falls for the immigrants' failure to find jobs.

30. Relatório. . . Ignácio Joaquim de Souza Leão. . . 16 de abril de 1888, pp. 14-15. The French version of José Osório de Cerqueira's pamphlet was

But the enthusiasm for immigration depended upon the immediate reactions to abolition, and official subsidization. In the early years of the Republic, the national government lost interest in the colonies; and the state government soon refused to bear the expenses. Equally important, the colonies themselves did not live up to either the immigrants' expectations or the promoters' promises. In late 1890 the immigrant shelter at Jaqueira housed 117 foreigners, mostly Italians, French, and Belgians, with a few Spaniards. Less than half were agricultural or day laborers; most were artisans, craftsmen, and tradesmen. On the Suassuna plantation, the immigrants complained that daily wages did not fulfill promises made them in Buenos Aires and Montevido, that food was insufficient, and that their skills were unnecessary for the work available. Fights broke out between immigrants and Brazilians who did not understand the French or Italian languages or customs, and many immigrants returned to Rio de Janeiro or Europe.[31]

La Province de Pernambuco au Brésil (Pernambuco, 1888). José Osório made a valiant attempt to sell Pernambuco: "In sum, the province of Pernambuco uncontestably has received a gentle salubrious climate; therefore, the adjustment there is easy for those born in other countries, who as farmers are looking for a place among very fertile lands populated by an hospitable people like ours. As a natural consequence, we say it is easy to build there the colonization centers, without which the development of resources and principally industries would be quite difficult." (*Ibid.*, p. 17.) For José Osório's plans, see "Reunião em palacio," *Diário de Pernambuco*, April 18, 1888. José Osório, "Publicações a Pedido: Colonisação sem colonos," *ibid.*, August 17, 1888. "Colonisação sem colonos," *Jornal do Recife*, August 14, 1888. "Criação de nucleos coloniaes," *Diário de Pernambuco*, April 22, 1888. 'Colonisação," and "Colonisação e Immigração," both in *ibid.*, May 5, 1888. "Sociedade Promotora de Colonisação e Immigração," *ibid.*, June 3, 22, 1888. 'Colonisação. Sociedade Promotora de Colonisação e Immigração de Pernambuco," *ibid.*, September 4, 1888. *Falla. . . 15 de setembro de 1888. . . Joaquim José de Oliveira Andrade*, p. 63. *Relatório. . . Joaquim José de Oliveira Andrade. . . 3 de janeiro de 1889*, p. 42. *Falla. . . 1 de Março de 1889 . . . Innocencio Marques de Araujo Góes*, p. 36. "Relatório com que o Excm. Barão de Souza Leão passou a administração da Provincia em 20 de Junho de 1889 ao Excm. 1º vice-presidente Barão de Caiará," *Diário de Pernambuco*, August 1, 1889. "Contractos para introdução de immigrantes," *ibid.*, June 29, 1889.

31. "Immigrantes," *ibid.*, September 27, 1890. "Delegacia da inspectoria geral das terras e colonisação," *ibid.*, October 18, 1890. "Jornal do Recife, A Immigração entre nós," *Jornal do Recife*, November 29, December 2, 6-7,

The national government made one last effort. It bought three more Jaboatão engenhos and renamed the enlarged colony, now totaling 2,200 hectares, Colônia Barão de Lucena. The colonists grew sugar cane, cocoa, and coffee, as well as food crops, on small lots up to 22 hectares. But the enlarged colony fared no better than its predecessors. The new immigration Inspector-General Manoel Barrata Góes agreed in 1893 to build an usina on a neighboring property, in partnership with the Portuguese financier Joaquim Lopes Machado, who had acquired lots in the colony. They formed the Companhia Progresso Colonial, capitalized at 400 contos, but the partners soon split. Barrata Góes accused Machado of nepotism and of trying to buy the entire colony. Machado retorted that the Inspector-General had defrauded the national treasury by overpaying for land and concealing the fact that he was the principal stockholder in the company. Colonists took sides in the dispute, and the ensuing scandal moved the national government to end its subsidy and cede the colony to state authorities. The incumbent governor, noting in disgust that the colony gave land to "wealthy people . . . merchants, capitalists, industrialists, and even civil servants, who could be everything but colonists or immigrants," cut off state support. In 1895 the colony was divided into lots and sold at public auction. The same governor also made a last modest attempt to promote immigration by sending agents to Spain and Portugal to contract for artisans, tailors, and stoneworkers, but the next governor rescinded the agreements.[32]

1890. "Hospedaria da Jaqueira," *Diário de Pernambuco,* November 30, 1890. "Excursão," *ibid.,* December 4, 1890. "Colônia Suassuna," *ibid.,* December 11, 1890. Paraizio de Valladares, "A emigração entre nós," *ibid.,* December 16, 1890. "A immigração entre nós," *Jornal do Recife,* January 8, 10, and 16, 1891; and *Diário de Pernambuco,* January 9, 1891. "Terras Publicas," *ibid.,* January 15, 1891. José Ottoni Ribeiro Franco, "O ex-encarregado da hospedaria ao publico," *ibid.,* January 18, 1891.

32. Gervasio Campello, "A extincta Colônia Suassuna," *ibid.,* June 11, 1892. J. Thiago da Fonseca, "Immigração para o norte," *ibid.,* June 18 and 22, 1892. Pereira da Costa, *Anais Pernambucanos,* v. IX, pp. 342-344. The Recife press published many articles by the disputants. See *Commercio de Pernambuco, Diário de Pernambuco,* and *Jornal do Recife,* between March 14 and August 31, 1894. Manoel Barrata Góes, *Nucleo Colonial Suassuna, O Delegado da Inspectoria Geral das Terras e Colonisação ao Excm. Sr. Governador do Estado e ao Publico* (Recife, 1894), is a collection of the immigration Inspector-General's

To their credit, Pernambucans rejected the worst of all possible immigration schemes, the importation of African contract laborers or Chinese coolies. The provincial assembly disapproved the idea of contracting free Africans, two of whom would be given to each slave-owner freeing a slave, in 1857. In the 1880's a planter spokesman again rejected the idea, and argued: "even if it were practical, it would be inconvenient and even immoral. Slow-moving and indolent, ethnologically speaking, the man of the black race in his native land is incapable of understanding a contract; he learns to work only from fear of punishment."[33]

Proposals to import Chinese coolies found equally little favor, mostly because the Pernambucans feared the Chinese would derive the greater benefits. The *Jornal do Recife* warned that the Chinese had "special qualities, their sobriety, economy, and dexterousness . . . assuring them, in all areas where they succeed in implanting themselves, the preference over more demanding national workers, and shortly thereafter the monopoly on trade and all the small industries and domestic services." The *Diário de Pernambuco* condemned the idea as "a new slavery disguised under the cloak of work contracts." Henrique Augusto Milet insisted that "there is no fusion possible between those races—Mongolian and Dravidian—with our race. The 100,000 to 200,000 Chinese today living in California and bordering states already are seriously compromising the future of that part of the U.S.A." While coolie labor had served sugar planters in Mauritius, Reunion, and other Antilles, Milet emphasized that the coolies had usually returned home after the harvest, neither spending their wages nor increasing the population in the host country.[34]

---

articles. *Mensagem. . . 1896. . . Alexandre José Barbosa Lima*, pp. 241-242. *Mensagem apresentada ao Congresso Legislativo do Estado em 6 de março de 1897 pelo Governador Dr. Joaquim Corrêa de Araujo*, pp. 58-59.

33. Cowper to Clarendon, Pernambuco, April 30, 1857, *Parliamentary Papers*, 1857-58, HCC, v. LXI, *AP*, v. XXIX, pp. 111-112. "Sociedade Auxiliadora da Agricultura de Pernambuco, Acta Provisória," *Diário de Pernambuco*, May 13, 1886.

34. "Trabalhadores asiaticos," *Jornal do Recife*, February 13, 1880. "Trabalhadores aziaticos," *Diário de Pernambuco*, March 16, 1880. Both these

Comparable census figures clearly demonstrate the failure of Pernambuco to attract agrarian immigrants. In 1872 the first national census counted 13,444 foreigners in Pernambuco; half were Portuguese and another 40 percent were Africans, both slave and free. In 1890 the foreign population numbered only 2,690, very likely an underenumeration but still quite low compared to the center-south states. In 1900 foreigners totaled nearly 11,000, and the Portuguese were 23 percent of this group (Tables 33 and 34). Those Portuguese mostly entered urban commerce and did not affect the labor supply to agriculture at all. In fact, the Brazilians so resented the Portuguese merchants that they had occasionally joined in revolt against them.[35]

Climate, land scarcity, and relative prosperity explain why immigrants did not go to Pernambuco but to the center-south areas. Pernambuco's climate may have discouraged Europeans. Most of the sugar zone lies no more than 400 meters above sea level and less than 10 degrees south of the equator. Temperatures during the summer harvest season average 81°F., and winter average temperatures remain high around 75°F. Periodic droughts devastated the sertão and the agreste, and although the zona da mata suffered less than those western regions, these natural disasters must have made interested Europeans hesitate. The SAAP believed that anthropologists and physiologists had proven that European colonists could not work in tropical agriculture. In São Paulo, on the other hand, the climate was more

articles criticized a pamphlet on the pros and cons of coolie labor by Salvador de Mendonça. Milet, *Auxilio à Lavoura e Crédito Real*, p. 35. *Idem.*, *A Lavoura da Canna de Assucar*, p. 15. SAAP, Livro de Atas no. 2, February 10, 1886, also published in *Diário de Pernambuco*, May 13, 1886. Even abolitionists used racist arguments against Chinese immigration. Toplin, *The Abolition of Slavery in Brazil*, p. 159.

35. The revolutions occurred in 1711, 1817, 1824, 1831, 1832-35, and 1848. The historiography of these movements is voluminous: see, for examples, articles listed in José Honório Rodrigues, *Índice anotado da Revista do Instituto Arquelógico, Histórico e Geográfico Pernambucano* (Recife, 1961). For convenient summaries, see Artur Cezar Ferreira Reis, "Inquietações no Norte," in Buarque de Holanda, *História Geral da Civilização Brasileira*, tomo I, v. 2; Quintas. "A agitação republicana no Nordeste," and Quintas, "O Nordeste, 1825-1850."

TABLE 34

PRINCIPAL IMMIGRANT GROUPS IN PERNAMBUCO

Year[a]	Sex	Portuguese	Italian	French	English	Spanish	African
1872	Male	5,637	292	201	179	189	3,065
	Female	1,009	35	91	38	10	2,349
	Total	6,646	327	292	217	199	5,414
1900	Male	2,015	397	56	146	124	
	Female	446	163	83	96	38	
	Total	2,461	560	139	242	162	n.a.

[a] The published 1890 census did not discriminate by nationality.

SOURCES: *Recenseamento da População . . . 1872*, v. XIII p. 218. Directoria Geral da Estatística, "Recenseamento da População em 31 de dezembro de 1900," pp. 142-143.

temperate. At an average elevation of between 800 and 900 meters, and lying more than 20 degrees south of the equator, summer temperatures were in the low 70's, below average winter temperatures in Pernambuco, and physical work was less exhausting.[36]

Many foreign visitors returned with favorable impressions of Pernambuco's climate because they never traveled outside Recife. They characterized the port as "remarkably salubrious," "the most delightful of any in Brazil," "the healthiest place on the coast," "the healthiest city in the tropics," and "one of the most, if not the most, healthful seacoast city in Brazil."[37] But Recife, constantly ventilated by year-round sea breezes, is not typical of the sugar zone. Foreign consuls with more experience than occasional visitors warned that "the European or Anglo-American would perish beneath the power of the sun in less than a week," "the European field or agricultural laborer could bear up against the climate but for a very short time," and "this is most decidedly not a country in any way suited for British immigrants." An interesting and important exception was an Italian consul, who told the

36. Manoel Correia de Andrade, *Paisagens e Problemas do Brasil* (São Paulo, 1968), p. 126. Mario da Veiga Cabral, *Coreografia do Brasil*, 28th edition (Rio de Janeiro, 1947), pp. 404, 490. Preston James, *Latin America*, 3rd edition (New York, 1959), pp. 471, 475.

37. Charles Waterton, esq., *Wanderings in South America, The North-West of the United States, and the Antilles, in the Years 1812, 1816, 1820, and 1824*, 2nd ed. (London, 1828), p. 93. Christopher Columbus Andrews, *Brazil, Its Conditions* (New York, 1887), pp. 111-112. Hastings Charles Dent, *A Year in Brazil* (London, 1886), pp. 246-247. Burke and Staples, Jr., *Business and Pleasure in Brazil*, p. 110. Report by Consul Atherton on the Commerce of Pernambuco for the Year 1882, U.S. Department of State, *Reports from the Consuls of the United States*, 1882-83, v. II, p. 317. Report by Edwin N. Gunsaulus, Pernambuco, September 28, 1900, *ibid.*, 1900, v. I, p. 754. The praise for Recife was not unanimous, however, An English visitor remarked that "the city of Pernambuco has little to recommend it to those who have no business dealings there," and an American naval officer reported "the dampness of the climate causes the town to look old, moldy, and decaying; the streets are narrow, filthy, and disagreeable." George Gardner, *Viagens no Brasil principalmente nas provincias do norte e nos distritos do ouro e do diamante durante os annos de 1836-1841*, translated by Albertino Pinheiro, (São Paulo, 1942, first published 1846), p. 65. Lieutenant Commander Henry Honeychurch Gorringe (USN), *The Coast of Brazil*, v. I, *From Cape Orange to Rio de Janeiro* (Washington, 1873), pp. 158-159.

SAAP that his homeland had "a numerous population, and its emigrants could also come to Pernambuco."[38]

But climate alone cannot be the best explanation for the lack of immigrants to Pernambuco, for when other compensations were present the European showed himself more than willing to live and work in the tropics. During the colonial period, for example, most European immigration before the eighteenth century settled in northeastern Brazil. Nor can one attribute the lack of immigrants to speculation, opportunism, political and religious discrimination, or slavery, for these factors were equally present in the center-south areas where immigration was massive. The abolition of slavery and the creation of a civil register of births and deaths in 1888 eliminated some barriers, and the separation of church and state and the religious freedom proclaimed in 1890 eliminated others.[39]

The scarcity of public lands in the zona da mata, the area of sugar production and that best served by public transportation,

38. In 1887, the sea breezes blew 93.7 percent of the time; the warmer land breezes blew 4 percent of the time; and calms prevailed only 2.3 percent of the time. "Regime dos Ventos," *Porto do Recife*, Anno 1, no. 1 (August, 1933). Cowper to Clarendon, Pernambuco, April 30, 1857, *Parliamentary Papers*, 1857-58, HCC, v. LXI, *AP*, v. XXIX, p. 111. Report by W. W. Stapp, Pernambuco, May 30, 1859, in U.S. Congress, *Executive Documents*, First Session, 36th Congress, v. 2, D. 4, p. 432. Stapp had reason to complain: he later died from disease in Recife, the fourth U.S. Consul to succumb on the job. Report by Henry F. Hitch, U.S. National Archives, Despatches from U.S. Consuls, v. 6. Doyle to Granville, Pernambuco, October 31, 1870, *Parliamentary Papers*, 1871, HCC, v. LXVIII, *AP*, v. XXXII, p. 106. Report by Bonham, Pernambuco, April 30, 1881, *ibid.*, 1881, HCC, v. XCI, *AP*, v. XXXV, p. 113. See also Report by Corfield, Pernambuco, June 30, 1875, *ibid.*, 1875, HCC, v. LXXVII, *AP*, v. XXXVI, p. 88. Consul Cohen to Marquis of Salisbury, Pernambuco, February 16, 1892, *ibid.*, 1892, HCC, v. LXXIX, *AP*, v. 7, p. 105. "Report on the Trade and Commerce of Pernambuco for the Year 1906," *ibid.*, 1908, HCC, v. CIX, *AP*, v. 48, p. 15. "Auxiliadora da Agricultura," *Diário de Pernambuco*, December 14, 1894.

39. Paes de Andrade, *Questões Econômicas*, pp. 55, 60. "A agricultura do norte e a colonização nacional," *Diário de Pernambuco*, June 11, 1887. Thomas Adamson, Jr., to William H. Seward, Pernambuco, October 23, 1866, U.S. National Archives, Despatches from U.S. Consuls, v. 8. *Relatório. . . Antonio Borges Leal Castello Branco* (1865), p. 60. For changes in laws, see Calogeras, *A History of Brazil*, p. 278, and Burns, *A Documentary History of Brazil*, pp. 288-289.

discouraged immigrants to Pernambuco. "As long as the lands in the coastal area, nearest the consumption centers, most populous and best endowed with transportation, are locked up by large land-owners who do not use them, nor cede them for farming . . . as long as we do not guarantee the immigrant the easy acquisition of the land surplus . . . we will never see the European exodus headed for Brazil," warned one provincial president. Another advised: "In this province, where the public lands are imbedded within the private domain and occupy small areas, it is difficult to think about creating colonies, due to the need to resort to expensive and somewhat risky purchases of the [privately] worked lands." If instead Pernambuco enjoyed the "great expanses of public lands which allow the creation of important colonies" in Rio Grande do Sul, Santa Catarina, and Paraná, immigrants would come. In São Paulo, land in the coffee region was also scarce, and immigrants began as salaried workers. But they could at least hope to earn enough to buy their own lands in the west, where new coffee plantations were gradually being created.[40]

Relatively low entry costs in the coffee industry also increased the possibility of the immigrant's one day becoming a landowner in Brazil. To prepare his product for export the coffee producer had to pick, shell, sort, clean, and bag—all purely manual or mechanical operations. Even if he completely mechanized the shelling, sorting, and cleaning, he would spend no more than one or two contos. The sugar producer, on the other hand, not only cut and ground his cane, but also transformed it chemically into sugar and alcohol. A substantial investment in machinery was

40. Falla . . . Diogo Velho Cavalcante de Albuquerque . . . 1 de março de 1871, p. 39. Falla . . . João Pedro Carvalho de Moraes . . . 1 de março de 1876, p. 82. See also Relatório . . . ACBP . . . 6 de agôsto de 1875, article xxvi. On the German colonies in Rio Grande do Sul, see Jean Roche, La colonisation allemande et le Rio Grande do Sul (Paris, 1959). Tereza Schorer Petrone, "Immigração assalariada,' in Buarque de Holanda, História Geral da Civilização Brasileira, tomo II, v. 3, pp. 291-293. Michael M. Hall, "The Origins of Mass Immigration in Brazil, 1871-1914," Ph.D. dissertation at Columbia University (New York, 1969), chapter IV. Idem., "The Italians in São Paulo, 1880-1920," paper presented to the American Historical Association annual meetings, New York, December 1971, p. 6.

inescapable, and modernization, as we have seen, cost 50 contos at least. We do not mean to suggest that Italians became coffee *fazendeiros* one generation after arrival in Brazil, for Michael M. Hall has shown that such success stories were hardly the rule. But São Paulo's immigration promoters could give more encouragement, and the immigrants had more to write home about, than did their counterparts in Pernambuco.[41]

Far more than climate or land scarcity, Pernambuco's relatively poor economy and prospects discouraged immigration. Between 1890 and 1910, the years of heaviest European immigration, coffee exports earned an annual average return between 400,000 and 500,000 contos, while sugar exports earned less than 50,000 contos, or one-twentieth as much income (see Table 2). The lucrative coffee economy permitted São Paulo governments to spend 1,000 contos per year to subsidize immigrants' transatlantic passage and initial expenses in Brazil; at this time Pernambuco officials were offering little or no aid to immigration promoters. The coffee boom also produced a slight wage differential favorable to São Paulo, where field hands earned 1$000 to 1$400 per day while comparable workers in Pernambuco earned only $800.[42] Those immigrants who became food growers for domestic markets in São Paulo also earned better incomes than their counterparts in Pernambuco's agreste and sertão, because the coffee

41. Milet, "A Colonisação, *Diário de Pernambuco*, May 17, 1888. Viotti da Costa, *Da Senzala à Colônia*, pp. 184-185. Celso Furtado also stresses the lower capital requirements of coffee, *The Economic Growth of Brazil*, p. 124. Hall, "The Origins of Mass Immigration," pp. 139-149. Other scholars have emphasized the immigrants' social mobility. See Dean, *The Industrialization of São Paulo*, p. 50. Thomas H. Holloway, "Condições do mercado de trabalho e organização do trabalho nas plantações na economia caféeira de São Paulo, 1885-1915. Uma análise preliminar," *Estudos Econômicos* (São Paulo), v. 2, no. 6 (December 1972), pp. 145-177.

42. Viotti da Costa, *Da Senzala à Colônia*, pp. 195-196. Our Table 32. Wage data for 1883. Both Pernambucans and foreigners complained about low wages. SAAP, *Trabalhos do Congresso Agrícola*, p. 311. "Representação," *Diário de Pernambuco*, August 13, 1889. Consul Cohen to Mr. Wyndham,. Pernambuco, February 5, 1892, and Hugh Wyndham to the Marquis of Salisbury, Rio de Janeiro, February 19, 1892, both in *Parliamentary Papers*, 1892, HCC, v. 70, AP, v. 32, pp. 112-113. Gonsalves de Mello, "Trabalhadores Belgas em Pernambuco," p. 17.

prosperity had generated greater demand and had encouraged investments in transportation which lowered freight costs.

This difference in relative prosperity naturally was reflected in production growth. Between 1876 and 1910 the average annual amount of sugar made in Pernambuco increased from 116,000 to 142,000 tons. (Table 5). During the same period the average annual volume of coffee exports increased from 220,000 to 827,000 tons (Table 2). This latter nearly four-fold increase required sizeable additions to the labor force in the coffee growing areas, such as were provided by immigrants, while the much smaller growth in the sugar industry could easily have occurred using the natural increase of the Brazilian population. In fact, the decline in wage levels in Pernambuco after 1870 suggests that the natural increase exceeded the demand for labor in the sugar industry. One should also remember that the modernization of that industry after 1870 increased productivity in the industrial sector through capital investments and further reduced the demand for labor.

Various modern scholars have argued that the difference in relative prosperity also reflected a difference in the basic nature of the productive systems in the northeast and the center-south; the former is usually characterized as feudal, and the latter, capitalist. [43] We have already seen in the previous chapter that supposed differences in the treatment of slaves have been attributed to the feudalism-capitalism dichotomy. In similar fashion Gilberto Freyre has argued that "in the north and northeast, the economy based on cane farming and sugar manufacture was developed on the basis of feudal relations between landowners and field slaves, so that the European colonists did not consider themselves strong enough to overcome [these relations], as long as the latifundia and monoculture endured, even without slavery."[44] Presumably Freyre is referring to the continuance of sharecropping and squatting arrangements, whereas the immigrants in the center-south

43. Viotti da Costa, *Da Senzala à Colônia*, p. 460. Genovese, *The World the Slaveholders Made*, p. 81. Ianni, *Raças e Classes Sociais no Brasil*, pp. 79-80.
44. Gilberto Freyre, *Ordem e Progresso*, v. 2, pp. 399-400.

rejected sharecropping and insisted on contracts and salaries.[45] But even in the coffee area, "the system which spread in the last decades [of the Second Empire] was less advantageous to them than sharecropping," and various contemporary observers reported "unhappy continuities between slavery and the new system."[46] Thus while relations between employers and rural workers in the center-south may have "modernized" vis-à-vis the northeast, it is not clear to what degree this brought benefits to the workers, and therefore it is even less clear that this difference, above and beyond the difference in material prosperity, constituted an attraction for immigrants.

Since the Pernambuco sugar planters did not need immigrants, we are not surprised to find that they hardly ever served as directors of immigration promotion societies, whereas in São Paulo the coffee planters frequently played important roles in such societies. The planters' indifference, and the province's failure to hold those few immigrants it had attracted, led many to disparage the whole idea. A provincial deputy wondered whether "from the realization of the desideratum of the proponents of foreign colonization, our agriculture will fall into the power of foreigners, as has already happened with our trade?" A planter regretted that "we are condemned to receive only immigrants with *white hair and blue eyes*, who, if they do us the favor of coming to our country, do so to *civilize* us, and soon become owners of the best lands at the state's expense." The *Diário de Pernambuco* warned against the danger of large homogenous colonies becoming "new States within the State."[47]

Other critics of immigration attacked the immigrants them-

45. Sérgio Buarque de Holanda, "As colônias de parceria," in *idem.*, *História Geral da Civilização Brasileira*, tomo II, v. 3, pp. 245-260. Viotti da Costa, *Da Senzala à Colônia*, pp. 188-202.

46. *Ibid.*, p. 199. Hall, "The Origins of Mass Immigration in Brazil," pp. 119-123.

47. Tavares de Mello, *O Brasil Agricola*, Anno IV, no. 23 (August 25, 1866), pp. 360-361. Um agricultor, "O abolicionismo e a lavoura," *Diário de Pernambuco*, April 6, 1884. "Diario de Pernambuco: A Colonisação," *ibid.*, October 3, 1896. See also Um agricultor, "A crise agricola," *ibid.*, May 8, 1897.

selves. The ACBP characterized the newcomers as "the scum of many European cities," and certain planters became indignant that "our government could spill gold from overflowing hands and it was only creating a miserable sewer for the garbage and excrement of foreign countries." Even the progressive Governor Barbosa Lima noted bitterly, "In this rush of fanatics, they do not see the moral and social evils and complications brought to the bosom of our people by the annual transporting of thousands of citizens of other countries, carriers of a thousand varieties of socialism and anarchy, goaded by the desire to get rich, irritated by discontent, and secretly molded by habits and customs which do not harmonize with our own." The governor suggested that since Pernambuco did not attract enough immigrants to warrant the full quota assigned by the national government as a subsidy, that money be paid to the state government, which would use it to hire unemployed Brazilians, "the best workers for the reconstruction of our industrial order."[48]

Despite complaints about the quality of workers and the momentary worries provoked by the gradual abolition laws, the planters in general could be more than satisfied with the existing supply of free labor. Squatters and sharecroppers required little cash outlay and could usually be compensated with the abundant unused land on the plantations. Wage workers demanded cash, but after the early 1870's the wage level fell steadily. All three types of free labor could be hired and fired at will, without the complications of contracts or indemnizations.

48. *Relatório da Direcção da Associação Comercial Beneficente de Pernambuco apresentado à Assembléa Geral da mesma em 6 de agôsto de 1868*, cited in Flávio Guerra, "Memórias de uma Associação (História do Comércio do Recife)," unpublished manuscript, 1965, pp. 40-41. "Sociedade Auxiliadora da Agricultura de Pernambuco—Acta provisoria da sessão extraordinária do conselho administrativo havida no dia 10 de fevereiro de 1886," *Diário de Pernambuco*, May 13, 1886. For similar sentiments see "Jornal do Recife," *Jornal do Recife*, December 2, 1890 ("For us, colonization is . . . expenses, no more") and Gomes, "Agricultura—A Lavoura e o Projeto N. 7," *Diário de Pernambuco*, April 23, 1893 (for Gomes the Suassuna colony was "a fountain of enormous waste of the people's sweat"). *Mensagem. . . 1893. . . Alexandre José Barbosa Lima*, pp. 73-75.

In comparing the welfare of the free worker with that of the slave, Gilberto Freyre has observed:

There is no doubt that there was more aid to the worker under the patriarchalism of the old engenhos than in the great majority of today's usinas. No one denies that there was strictness and even cruelty in the exploitation of the slave by the white man of the Big House. The common occurrence, however, was the master protecting the Negro of the slave-quarters better than the usineiro of today provides for his worker. The masters even kept the old or sick Negroes on the plantations at their own expense. In the majority of the old engenhos, life passed more sweetly and more humanely for everybody than on the usinas.[49]

Such romantic nostalgia may be cloying, and our data hardly confirm the sweetness of the slave's lot. But one cannot escape the conclusion that the free rural laborer in the later nineteenth century enjoyed little material advantage over the slave. His diet was virtually identical, his job tenure was less secure, and his rewards were paltry unless he had some particular skill. Of course he enjoyed the freedom to choose where and when he would work, and probably he was not as frequently subject to corporal punishment as the slave. But these advantages may not have compensated for his deteriorating standard of living. In the transition from slave to free labor, the sugar planters seem to have derived the greatest benefits, and the workers the least.[50]

49. Freyre's preface to Bello, *Memórias de um Senhor de Engenho*, p. x. See also Jovino da Raiz, "O trabalhador negro no tempo do bangué comparado com o trabalhador negro no tempo das uzinas de assucar," *Estudos Afro-Brasileiros, Trabalhos apresentados ao 1º Congresso Afro-Brasileiro, reunido no Recife em 1934* (Rio de Janeiro, 1935), pp. 191-194; and Manuel Diegues Júnior, "O Bangué em Pernambuco no Século XIX," *Revista do Arquivo Público* (Recife), Anos VII-X, nos. IX-XII (1952-1956), pp. 17, 29.

50. One of the first modern writers to reach this conclusion was Celso Furtado, *The Economic Growth of Brazil*, p. 152.

9

CONCLUSIONS

*I*n the late nineteenth century, two crisis disturbed the Brazilian sugar economy. One arose in the market when beet sugar competition in Europe usurped traditional cane sugar markets. The other was a serious social crisis that arose within Brazil when the imperial government moved to abolish slavery. The Brazilian planters attempted capital improvements and reorganization of production to cope with the market crisis, but failed. The only real solution would have been integration into a northern hemisphere market via recolonization, but that would have entailed heavy political costs. The planters coped better with the social crisis. They succeeded in transferring the losses suffered in export markets to the plantation work force in the form of depressed wages and poor working conditions. Their efforts, aided with government subsidies, perpetuated their dominance in Brazil's sugar areas. Thus "modernization," understood as technological advance and the abolition of forced labor, failed to produce real changes.

The market crisis affected all cane sugar producers, who were excluded from about half the world market by century's end. The shift away from cane sugar consumption did not reflect changing tastes, for cane sugar and beet sugar for most purposes were interchangeable. In fact, the volume of world cane sugar production increased fivefold during the late nineteenth century. Nor did the shift indicate the demand was declining, for average annual per capita consumption in England rose from 15 pounds in the early 1800's to 72 pounds by the later 1880's.[1] Such accelerating consumption should have benefited suppliers. In general, it did; but not Brazil.

The crisis first appeared in falling prices in the early century; Brazil's sugar export revenues began serious slipping in the 1860's and 1870's. Although export volumes continued to grow, and the trade recovered briefly in the 1880's, in subsequent years Brazil suffered disastrous decreases in both revenues and volumes.[2] While the beet sugar competition affected all cane sugar producers, and export growth in Martinique, Guadeloupe and Mauritius also declined, it was not generic. Cuba's successful experience demonstrates that Brazil suffered from more specific disadvantages. To reformulate my interpretation of these difficulties, therefore, I shall briefly compare the Cuban sugar economy with that of Pernambuco, the northeastern province which led Brazil in sugar exports throughout the late nineteenth century.

A glance at the map suggests immediately that Cuba's principal advantage over Brazil, as far as selling sugar to large consumer markets, was its proximity to the United States. Certainly the island's location allowed lower transportation costs to the United States than those charged on sugar shipped from Brazil, or virtually any other foreign supplier. As ocean transport costs steadily declined in the nineteenth century however, because of steam engines and other advances, Cuba's advantage over Brazil in this regard became less crucial.[3] But Brazil still failed to gain perma-

1. Hermann Paasche, *Zuckerindustrie und Zuckerhandel der Welt* (Jena, 1891), pp. 411-412. I averaged consumption data for 1801-5 and 1886-90.

2. World War I interrupted European beet sugar production and thereby stimulated all cane sugar producers. After the war, however, Western Hemisphere cane sugar producers underwent recession. While Cuba recovered in the later 1920's, Brazil did not, and by the 1930's the Brazilian sugar industry was becoming more and more dependent on government aid. "O Açúcar na vida econômica do Brasil," pp. 233-235. Cuban Economic Research Project, *A Study on Cuba* (Miami, 1965), p. 235. Barbosa Lima Sobrinho, *Problemas Econômicos e Sociais da Lavoura Canavieira*, p. 33. Singer, *Desenvolvimento Econômico e Evolução Urbana*, pp. 324-325.

3. Douglass C. North, *Growth and Welfare in the United States* (Englewood Cliffs, N.J., 1966), pp. 109-110. North shows that an index of U.S. export freight rates, presumably roughly proportional to rates charged on ships bound

nent access to the rich North American market. In the 1880's, when Cuba was recovering from the Ten Years' War and low sugar prices forced many marginal mills out of production, Cuban exports declined, and Brazil tripled its sugar shipments to the United States. The success raised hopes of replacing lost European markets. Optimism swelled when the United States promulgated the McKinley tariff in 1890, which like the British Sugar Act of 1846 reduced charges on raw sugar and molasses. Brazil had gained strong access to the English market after 1846, so sugar exporters hoped for similar access to the North American market after 1890. In 1891 the United States signed a reciprocal trade agreement with Brazil; but all hopes were dashed when, in the same year, the United States signed the Foster-Canovas treaty with Spain, which extended the same favors to Cuba and Puerto Rico. After those islands separated from Spain in the late 1890's, the United States signed new reciprocal trade treaties which ratified their status as sugar colonies of the North Americans.[4]

Geographical proximity gave Cuba a favored position, and colonial preference agreements formalized the relationship. But these factors alone do not explain Cuba's access to the United States market, nor Brazil's exclusion. Economic considerations such as the supply of the crucial productive factors of land, labor and capital are more important.

Sidney W. Mintz has pointed to soil fertility as the determining factor in the sequential rise of sugar colonies in the Caribbean.[5] Many writers have commented on the excellence of Cuban soils for growing sugar, and this great fertility certainly contributed to high yields and Cuba's comparative advantage vis-à-vis

for the U.S., fell over 75 percent between 1815 and 1850, and again by nearly 50 percent between 1879 and 1908.

4. Hugh Thomas, *Cuba: The Pursuit of Freedom* (New York, 1971), pp. 288-289, 290-291, 457, reviews the agreements of the 1890's. Cuban Economic Research Group, *A Study on Cuba*, pp. 218-219, summarizes the Reciprocity Treaty of 1902.

5. Sidney W. Mintz, "Labor and Sugar in Puerto Rico and Jamaica, 1800-1850," *Comparative Studies in Society and History*, v. I, no. 3 (March 1959), pp. 273-280, reprinted in Foner and Genovese, *Slavery in the New World*, pp. 170-177, especially p. 171.

Brazil.[6] New Cuban cane lands produced as much as 119 tons of cane per hectare, and median lands in the 1870's yielded nearly 70 tons per hectare, whereas reported yields in Pernambuco never exceeded 60 tons per hectare.[7] As a result, Cuba could produce over one million tons of sugar from about 814,000 hectares of cane in the 1890's, when Brazil was getting no more than 100,000 tons of sugar from 417,000 hectares.[8]

While good land was abundant in Cuba, labor was relatively scarce. Unlike Brazil, which had exported sugar since the sixteenth century, Cuba only entered large-scale sugar production in the early nineteenth century as a consequence of Haiti's withdrawal from the world market. In the eighteenth century, Cuba had exported principally tobacco, but the volume of this activity had never reached such proportions as to cause massive importations of African slaves. As a result, when Cubans began to produce sugar, slaves were scarce and expensive.[9] The rapid expansion of Cuban sugar production outstripped the importation of African slaves, and by the 1840's Cubans were contracting Chinese coolie labor for the plantations.[10]

In Brazil's sugar areas, on the other hand, labor abounded. Whereas Cuba had fewer than 300,000 slaves in 1871, Brazil registered 1.5 million in 1873. Of course Brazil's slaves did not all work in sugar production, but just in Pernambuco, Alagoas, Sergipe and Bahia, the northeast's principal sugar exporting provinces, there were over 300,000 slaves.[11] Even when Brazil's

6. See, for example, Robert P. Porter, *Industrial Cuba* (New York, 1899), p. 282.

7. Moreno Fraginals, *El Ingenio*, pp. 94-96. I assume the Cuban *caballeria* of 33.3 acres equaled 13.5 hectares, after Ely, *Cuando Reinaba Su Majestad El Azúcar*, p. 438.

8. Porter, *Industrial Cuba*, pp. 281-282. In deducing the Brazilian area under cultivation, I assume yields of 60 tons per hectare and no more than 8 percent sugar extracted from cane. Probably median extraction yields were much lower, which would increase the total area under cultivation and aggravate the discrepancy with Cuba.

9. Franklin W. Knight, *Slave Society in Cuba During the Nineteenth Century* (Madison, Wisc., 1970), pp. 4-6, 29.

10. Duvon Clough Corbitt, *A Study of the Chinese in Cuba, 1847-1947* (Wilmore, Kentucky, 1971), pp. 1-26.

11. Knight, *Slave Society in Cuba*, p. 63. Toplin, *The Abolition of Slavery in Brazil*, p. 268.

sugar areas converted from slave to free labor, the number of native workers more than met the industry's demand, and real wages actually declined after 1870. Comparable wage data for Cuba are not available, but it seems clear that the labor scarcity situation continued in the late nineteenth century. Cuban workers began organizing as early as the 1850's, and strikes occurred after the 1860's, at least 30 years before comparable labor organization and activity in Pernambuco.[12]

If labor was scarce in Cuba, capital was not. Spanish, Spanish-American and Cuban capitalists financed the early sugar plantations at monthly rates exceeding 1.5 percent. Attracted by the island's proximity and economic prospects, United States capitalists also played a growing role in financing the plantations. When crisis struck, as during the Ten Years' War, the foreign capitalists became plantation owners both by necessity and by choice. The growing United States interest also took the form of annexationist movements in the 1840's through the 1870's.[13]

The happy combination of fertile land, scarce labor, and available capital allowed Cuba to lead the world in modernizing its cane sugar industry. By the 1860's, 70 percent of Cuba's 1,350 mills used steam engines, in comparison with only 2 percent of Pernambucan mills; as late as 1914 only one-third of Pernambuco's mills used steam power.[14]

But modernization meant more than just steam engines and multiple-effect vacuum pans. The growing size of sugar mills encouraged division of labor between the agricultural, cane-growing sector, and the industrial, sugar-making sector, because the mills simply needed more cane than any one plantation could supply. Moreover, if independent planters supplied the cane, millers could concentrate attention and investments on expensive machinery. In Cuba, this reorganization of production replaced the traditional mills (ingenios) with large capacity central mills.

12. Foner, A History of Cuba, v. 2, pp. 136-148.
13. See Ely, Cuando Reinaba Su Majestad El Azúcar, pp. 238-415 for descriptions of capitalists who financed Cuban sugar, and Thomas, Cuba, p. 271, for interest rates. On U.S. capital and annexationists, see Leland Jenks, Our Cuban Colony (New York, 1928), pp. 33-35, and Thomas, Cuba, pp. 207, 263, 271-275. Good prospects did not accompany comparable interest rates in Brazil.
14. Deerr, The History of Sugar, v. 1, p. 130.

The cane suppliers, known as *colonos*, included two categories: the independent grower, who formerly may have operated a mill but now only grew cane on his own land; and the dependent grower on central mill land. The Cubans experimented initially with contracted cane prices, but this system proved disastrous when sugar prices fell unexpectedly. The Cuban central owners then paid for cane on the basis of the current sugar price, and the system succeeded. United States investors' growing interest in Cuba provided much of the capital necessary to transform or create central mills. By the early twentieth century, 170 to 180 central mills completely monopolized Cuban production. The Ten Years' War and the Wars for Independence (1895-98) greatly accelerated depreciation of older mills, and thereby sped the disappearance of the traditional mill.[15]

In Brazil, private capital showed less interest in sugar, and technological modernization proceeded hesitatingly. The British appear to Richard Graham and Eugene Genovese to have played a role similar to that of the North Americans in Cuba, but the appearance is deceptive. Brazilian correspondents borrowed capital from British commercial banks in Recife and then made short-term loans at 1.5 percent monthly or more to the planters to cover operating expenses. But Brazil's comparative disadvantages in sugar production did not attract the same quantity of foreign capital as did Cuba. To effect the transformation of traditional mills to central mills, considerable government aid was necessary.

Official aid became available in the 1880's and 1890's and was essential for modernization. At first, the imperial government guaranteed profits to companies building central mills. These mills owned no cane fields and bought their raw material from independent suppliers. But two difficulties plagued the subsidized central mills. First, many of the foreign concessionaires perpetrated frauds, indicating they had little interest or faith in Brazil's sugar export markets. Second, the mills which reached completion had trouble regulating the cane supply.

15. Jenks, *Our Cuban Colony*, pp. 31-33; Thomas, *Cuba*, pp. 276-277. Cuban Economic Research Group, *A Study on Cuba*, pp. 92, 96, 235.

Cane growers maintained their own mills and refused to send cane if they did not like prices or conditions; they also resented being reduced to the status of sharecroppers. No exogenous forces such as war obliged the traditional mill owners to collaborate with the new central mills.

The Brazilians overcame these difficulties by relying almost exclusively on generous official loans to native entrepreneurs; the new republic founded in 1889 allowed the states to collect export taxes, and in Pernambuco these taxes were used to finance long-term low-interest mortgages. Moreover, the new modern usinas not only bought cane from independent growers but also owned their own cane lands to insure steady supply and independence from the still-powerful senhores de engenho. Thus the usina recreated to a large extent the productive organization of the traditional mill.

Some 60 modern mills were operating in Pernambuco by 1910, of which two-thirds had received subsidies. But another 2,000 traditional mills continued to supply local demand for crude sugar and cheap rum. Thus while government aid permitted modernization of a small portion of the mills, the transformation and reorganization were never completed.[16]

Even within Brazil, where domestic demand might have partly compensated for loss of foreign markets, distance and domestic competition limited the northeastern sugar industry's marketing possibilities. The Pernambuco sugar exporters attempted in the first decade of the twentieth century to cartelize the national market. Their attempts failed because the most populous consumer areas of Minas Gerais, Rio de Janeiro, and São Paulo had their own local sugar industries since the colonial period, and those industries were reluctant to join the cartel. The

16. Some contemporary observers believed that Brazilian sugar was inferior in quality. I am not sure to what extent this was the case, nor if true to what extent the inferiority could be ascribed to lack of capital investments to improve cane and its manufacture. For a sample criticism of Pernambuco sugar, see A. F. Howard, "Report for the Year 1896 on the Trade etc. of the Consular District of Pernambuco, *Parliamentary Papers*, 1897, HCC, v. 79; *AP*, v. 28, p. 10.

refiners in those areas, moreover, enjoying near monopsony because of the export crisis, played producers off against each other and frustrated the cartel's efforts.

Nathaniel Leff has recently argued that in Brazil the sugar producing areas' difficulties were further aggravated by the presence of the coffee economy. Since for all practical purposes the reigning foreign exchange rate for both sugar and coffee areas was the same, the flourishing coffee economy's high export earnings kept the exchange rate for sugar earnings above the level it otherwise would have found. Had the rate fallen faster, as would have happened if not buoyed up by coffee's earnings, then sugar exporters receiving foreign currency would have been able to buy more mil-réis, and thereby could have cushioned the export decline.[17] Such revenue effects are only clearly positive, however, when inflation does not affect domestic prices for exporters and when their demand for imports is low. Sugar exporters could have escaped the effects of a more acute domestic inflation, for market imperfections would have arrested the spiral's climb and the working class lacked bargaining power to raise wages. But the need to import expensive capital equipment for modernization would have eaten into these short-run revenue gains. Those gains would not have been sufficient to stimulate the growth of a local capital goods industry, as Celso Furtado has suggested, and thereby reduce the dependence on foreign imports, for even in São Paulo, where economic conditions were much more propitious, such an industry did not appear until the 1930's. Previous efforts in that direction had depended far more upon protective tariff legislation and trade treaties than upon exchange rates.[18]

The sugar export market difficulties led to regional differences within Brazil, and many Pernambucans blamed their problems on political discrimination. But political discrimination was

17. Nathaniel Leff, "Economic Development and Regional Inequality: Origin of the Brazilian Case," *Quarterly Journal of Economics*, v. LXXXVI (May 1972), pp. 258-259.

18. Nathaniel Leff, *The Brazilian Capital Goods Industry* (Cambridge, Mass., 1968), pp. 8-40. Furtado, *The Economic Growth of Brazil*, p. 224.

a consequence, not a cause of Pernambuco's difficulties. Pernambuco received the largest imperial subsidies of any province for central mill building of any province. The European immigration and early industrialization which Rio de Janeiro and São Paulo enjoyed resulted from the strength of those areas' economies far more than any imperial subsidies. In the republic, moreover, when northeasterners in general disappeared from the top levels of the executive, charges of regional discrimination continued but with no greater justification. State budgets in both the northeast and the center-south funded the principal economic development programs such as usinas and immigration; Pernambuco could hardly blame São Paulo for not spending money outside the state line. Even in the twentieth century, when federal agencies such as the *Superintendência para o Desenvolvimento do Nordeste* (SUDENE), the *Banco do Nordeste* and the income tax bureau encourage industrial investment in the northeast, the meager results provoke continued complaints about the growing gap between the northeastern economy and that of the center-south.[19] But the basic fault, to paraphrase Cassius to Brutus, lay not in São Paulo but in the northeast itself. Internal colonization may have aggravated the differences between the areas, as when coffee area politicians excluded northeasterners from influential posts in the national government, as Andre Gunder Frank has written, but that difference derived mainly from the contrasting position of Brazilian coffee and sugar in the world markets.[20]

If my analysis is correct, the only possible solutions to Pernambuco's sugar market crisis were integration into a northern hemisphere market via recolonization: new use of the productive resources employed in sugar; or continuation with the mainte-

19. For an early optimistic study of such programs, see Stefan H. Robock, *Brazil's Developing Northeast: A Study of Regional Planning and Foreign Aid* (Washington, 1963). For later more pessimistic accounts, see Alberto Tamer, *O Mesmo Nordeste* (São Paulo, 1968) and Riordan Roett, *The Politics of Foreign Aid in the Brazilian Northeast* (Nashville, 1972).

20 Gunder Frank, *Capitalism and Underdevelopment in Latin America*, p. 170. Gadiel Perucci, "Le Pernambouc (1899-1930)," chapter 1, stresses internal colonization.

nance of the status quo and transferal of the cost to other groups in Brazilian society.

Had the northeast seceded from the rest of Brazil, as Nathaniel Leff speculates and as was actually attempted in 1824, the sugar exporters undoubtedly would have experienced a faster falling exchange rate, and short-run revenue gains. Secession would have freed the northeast from the pernicious effects of the center-south's regional preponderance. It might also have inhibited labor mobility and thereby depressed the cost of labor whether slave or free. Moreover, it would have ended the drain of capital to the south in the form of imperial taxes, and facilitated local subsidies for modernization. But none of these changes would have restored Pernambuco's export markets, and the small local population could not have absorbed the area's production.[21]

Only if the independent sugar-producing region joined the formal British Empire or the informal United States empire, might it have fared better in the international market. If the independent northeast could have offered especially favorable conditions to attract foreign capital, the home country subsequently might have made preferential trading arrangements to guarantee markets to the colony. Of course such neocolonialism would have entailed high political costs in terms of loss of sovereignty, as Cuba experienced in the first half of the twentieth century; nevertheless, the sugar exporters would have improved their earnings.[22]

A land reform that redistributed resources employed in sugar production would not have salvaged the private economy, for no other single export crop was as lucrative as sugar in Pernambuco. At times cotton drew investments from sugar areas, but cotton's position on Brazil's export list depended upon the temporary absence of leading world producers from the international market. Once those producers returned, as after the United States War for Independence, the Napoleonic Wars, and the United

21. Leff, "Economic Development and Regional Inequality," p. 259.
22. Nathaniel Leff has considered the advantages of neo-colonialism for Brazil. "Economic Retardation in Nineteenth Century Brazil," *Economic History Review*, 2nd series, v. XXV, no. 3 (August 1972), p. 505.

States Civil War, Brazil was displaced from world cotton markets. Tobacco had been grown near the sugar zone in Bahia's Recôncavo, but as an export crop it was never as rewarding as sugar. Cattle-raising had traditionally enlivened the economy in Pernambuco's backlands, but that area lacked good pasturage and permanent water supply while Minas Gerais and Rio Grande do Sul, which were nearer the populous center-south and were rich in cattle and dairy farming, dominated the national market. Finally, coffee, cocoa and natural rubber became important exports in the late nineteenth century, but the climates and soils required to grow these crops were not found in Pernambuco. In the absence of any alternative export activity yielding returns equal to sugar, therefore, the planters' concentration on sugar is understandable.

Since neither regional independence nor different use of land were real possibilities, it is not surprising that the planters chose to meet the export crisis with technological modernization. The government subsidies built a protected competitive position for the usina owners and encouraged investments in plantation. The failure to cartelize the national market prevented the planters from recovering losses through higher prices to domestic consumers. The planters were able, however, to pass the costs of falling export markets to the workers in the form of depressed wages, poor conditions, and insecure job tenure.

As world sugar competition created a crisis in Brazil's export markets in the late nineteenth century, the abolition of slavery, begun with the ending of the African trade in 1850, created a social crisis at home. The timing of abolition resulted from political pressures as well as from fluctuations in export earnings; but it directly affected production costs and the organization of the sugar industry's labor force. Various regions of Brazil reacted differently to abolition, depending upon the value and need for slave labor.

In Pernambuco the planters' monopoly of land in the sugar zone, and sugar's continued comparative advantage within the province gave them economic and political supremacy. This power enabled them to convert from slave to free labor with a minimum of inconvenience, and thereby pass much of the cost of the export crisis to the free workers, who had virtually no bargaining power at all.

The number of mills actually increased in the later nineteenth century. If not a product of sugar's comparative advantage, one might ascribe the new investment to noneconomic motives such as prestige and political influence associated with large land holdings. Some planters in fact may have acquired land for such reasons, for the absence of real estate taxes or other carrying charges left landholding virtually costless. But the sugar oligarchy's positive pressures for modernization show they were fully capable of pursuing their economic self-interest. Sugar remained the preferred investment; and new mills made money.

While the planters worked only very small portions of their estates, the size of their holdings effectively excluded the development of a small farmer class which might have challenged their supremacy. Planters monopolized local political office, where their power reflected their monopoly on the means of production. To be sure, planters differed among themselves and fought out these differences in the imperial political parties of Liberals and Conservatives, and also in and out of courts, but these conflicts never spread far enough to endanger the class's supremacy. Even when popular elements and urban groups entered the fray, the planters continued to participate on both sides, and the defeat of individuals never signified the class's defeat. Thus the political supremacy enjoyed since the early days of colony continued into the twentieth century. Office did not always entail power, but in rural Pernambuco, at least, the planters both reigned and ruled.

This power enabled the planters to benefit from the transition from slave to free labor. After 1850 the outlawed African slave trade effectively stopped. But the flourishing coffee industry in the center-south still relied on slave labor, and its demand forced prices to climb rapidly. These rising prices, in conjunction with the sluggish sugar export trade, made slave labor in sugar less and less profitable. Even without working out formulas for capitalization of future income streams realizable through the use of slave or free labor, it is clear that the prevailing high interest rates and sugar's uncertain markets would have obliged planters to discount heavily future earnings from slave labor.

In fact, the planters converted from predominantly slave to predominantly free labor by 1872, when the first general census in imperial Brazil showed a preponderance of free agricultural workers in all Pernambuco's sugar counties. The conversion to free labor proceeded in several ways. Coffee's demand for slaves led to an interprovincial slave trade, which drained slaves from the northeastern sugar areas to the center-south. Other Pernambuco slaves simply died. A fair number were emancipated prior to final abolition in 1888, through legal measures, private philanthropy, and manumission.

Declining profitability accounts for the sugar planters' quick conversion to free labor. The planters simply could not afford to purchase slave labor. Nor did they attempt to breed their own slaves. The imputation that such an industry existed in the United States has been based on demographic data and contemporary testimony. While some scattered references to such activities in the coffee areas of Brazil do exist, the data do not support a similar imputation in Pernambuco.[23] Three indicators for Pernambuco imply the existence of a slave-breeding industry, but when other factors are considered the case is not very convincing.[24] First, decreasing male-female ratios suggest a growing interest in women for breeding purposes. But the trend is better explained by the fact that after 1850 the absence of sex-selective African imports and the trading south of prime males allowed the sex ratio to return to natural proportions. Second, the median age of slaves in Pernambuco increased and was higher than that in São Paulo after 1872; this change probably reflects the northeast's selling young adults to the center-south, but not necessarily breeding slaves for that purpose.[25] Finally, the increasing fertility ratio of slave children per 1,000 slave women of child-bearing age might also reflect deliberate breeding. More likely,

23. See Toplin, *The Abolition of Slavery in Brazil*, p. 19, for evidence that breeding occurred. Conrad, *The Destruction of Brazilian Slavery*, pp. 31-32 denies breeding occurred.

24. My Table 24.

25. *Recenseamento* . . . *1872*, v. 19, p. 433. Slave median age in Rio de Janeiro, however, fell within the same cohort as in Pernambuco, 26 to 30 years. *Ibid.*, v. 15, p. 358.

however, the increase corresponds only to the trend in the general population and improved recording. At best, a slave-breeding industry could have been profitable for only the generation between 1850 and 1871, when the interprovincial trade reached its height and before children born of slaves were declared free.

The conversion to free labor also resulted from the nature of master-slave relations. Technological improvements usually entailed complicated machinery. The masters refused to invest in the slaves' education to enable them to operate this machinery; and they also feared to trust slaves with such expensive equipment, for sabotage was common.

The northeastern planters clearly treated slaves as labor inputs. Thus assertions that planters in the aggregate treated slaves as members of an extended family, or a quasi-feudal lower estate, do not apply in Pernambuco. Gilberto Freyre, more than anyone else, has elaborated the view that the slaveowner's paternalistic affections for their slaves, and in particular their promiscuity with female domestics, softened slavery. Since planters often freed their illegitimate children, the conversion to free labor may have resulted more from humanitarian motives. No matter how widely such attitudes were shared, however, the natural increase of mulattoes never met the demand for labor, and over half the Pernambuco slave labor force had to be imported from Africa before 1850. Moreover in the post-1850 period, only 10 percent of the slave population became free workers in Pernambuco before 1888. Masters on the whole were probably indifferent as to whether their work force contained ex-slaves or people born free. When they freed slaves with the qualification that they continue working on the plantations, they were guaranteeing their labor supply as much as protecting the extended family.

By the same token, masters were willing to forego the social prestige and power based on slave-ownership. While traditionally they may have considered their slaves primarily as a lower estate in a quasi-feudal society, after 1850 the social value of slavery fast receded into the background, and the planters replaced slaves with nearly as subservient client populations of squatters, wage workers, and sharecroppers. Thus the planters protected their status while abandoning the slave estate.[26]

My emphasis on economically-motivated gradual abolition in Pernambuco disputes Robert B. Toplin's assertion that slavery in Brazil was only abolished after increased abolitionist activity and violence by slaves themselves. Toplin accepts at face value the slave population reports of the later 1870's and early 1880's, which I have shown to be suspect and upon which Toplin himself casts doubt. Counting as slaves the 27,000 ingenuos living in Pernambuco after 1871, as does Toplin, might more accurately reflect social reality, but certainly not the economic importance of slavery, for the oldest of these children would barely have reached working age by 1888, the date of final abolition. Including ingenuos would raise Pernambuco's total reported slave population for 1887 to 70,000, still less than one-half of the total in 1850. While Toplin's argument may hold for the center-south, and he documents well the frequency of mass escapes in São Paulo, there simply is no evidence that some 39,000 slaves fled their owners in Pernambuco between 1886 and 1887; evidence of mass flights does exist for early 1888 but is not pertinent.[27]

Slavery became so unprofitable that certain particularly poor northern provinces, such as Ceará and Amazonas, actually abolished slavery in the early 1880's, several years before national abolition. Other areas like Pernambuco phased the institution out in practice if not in law. Thus where economic activities using slave labor enjoyed strong markets, as in Brazil's center-south or the United States cotton states, slavery remained profitable and

26. For a recent example of this Weberian analysis, see Florestan Fernandes, *Comunidade e Sociedade no Brasil* (São Paulo, 1972), pp. 309-314, 399-404, and Toplin, *The Abolition of Slavery in Brazil*, p. 13. Elsewhere in Brazil planters allegedly kept slaves from force of habit, despite declining profitability. Stein, *Vassouras*, pp. 229-230. Eugene Genovese has maintained that planters owned slaves in the American south for a variety of noneconomic reasons. *The Political Economy of Slavery: Studies in the Economy and Society of the Slave South* (New York, 1961), pp. 30, 50.

27. Toplin, *The Abolition of Slavery in Brazil*, pp. 20-24, 94, 109, 223.

violence became necessary to abolish it. Where such activities did not boom, slavery more quickly became a liability once the international traffic stopped; and the slaveowners themselves took initiatives to substitute free labor.[28]

If the slaveowners treated their slaves as productive factors, whose cost determined whether they or free workers would be used, then clearly capitalist production used slave labor. As Sidney W. Mintz has written, "The slave plantation, producing some basic commodity for the mother country, was a special, emergent capitalist form of industrial organization."[29] If one denies that capitalism could use slave labor, then he may exaggerate the importance of the transition to free labor. Thus Eugene D. Genovese, for whom capitalism is "the mode of production characterized by wage labor and the separation of the labor force from the means of production—that is . . . labor power itself has become a commodity," sees in Brazil's northeast a transition from slavery to seigneurialism with capitalist elements represented by the increasing importance of wage and salary workers. But the transition in Pernambuco did not entail any major changes in the relationship between the workers and the means of production, nor, for that matter, between the workers and the distribution of production. Rural wages yielded little more claim on land, capital, or the fruits of production than had slaves' rations. At best, free workers had a greater claim over their own labor power, since they could refuse to sell it, but eventually they would face the choices of stealing or starvation. Our examination of free labor conditions in Pernambuco after 1850 makes clear that abolition in the long run meant little.[30]

28. On the profitability of slavery in the U.S. cotton south, see Robert William Fogel and Stanley L. Engerman, "The Economics of Slavery," in *idem.*, *The Reinterpretation of American Economic History* (New York, 1971), pp. 311-341.

29. Sidney W. Mintz, "Review of Stanley M. Elkins' Slavery," *American Anthropologist*, v. LXIII, (June 1961), pp. 579-587, reprinted as "Slavery and Emergent Capitalism," in Foner and Genovese, *Slavery in the New World*, pp. 27-37.

30. Genovese, *The World the Slaveholders Made*, pp. 16, 90-91. Both Marx and Engels considered slave labor to be distinct as a mode of production and a form of exploitation, along with serf labor and wage labor. Karl Marx, *Capital*, translated by Samuel Moore and Edward Aveling, and edited by

Free labor on the sugar plantations entailed wage and salary work, squatting, and sharecropping. Wages in the sugar zone increased in the later 1860's and early 1870's because of the increased demand by local railroad builders and cotton planters. But by the last quarter of the century the growing numbers of free men and the deteriorating export position placed strong downward pressure on wages. At the same time basic food prices of manioc flour, beans and jerked meat rose, with the result that real income fell even faster than wages. While some workers may have grown their own manioc and beans on squatters' plots, few kept cattle for meat. Those who depended exclusively on wages, moreover, bought all staples and suffered acutely from the falling purchasing power. Sharecropping, on the other hand, demanded at least tools and draft animals. But the success story of José Marreira, who in José Lins do Rego's novel *Banguê* rose from sharecropper to senhor de engenho, was atypical. Relative to the rural work force, only a few individuals were sharecroppers, and only a tiny fraction of these accumulated the capital necessary for landowning.[31]

None of these labor modes entailed much bargaining power. Planters could hire and fire wage workers and squatters at will in Pernambuco's labor-abundant economy, and frequently they made demeaning social or political demands on workers and their families which could not be refused. Even sharecroppers, who by virtue of their modest capital holdings qualified as a kind of rural middle class, could be dispensed with literally overnight, for the planters' monopoly on land meant there were always plenty of landless workers willing to grow cane for a share of the sugar produced.

Frederick Engels (New York, 1906, first published 1867), p. 591. Engels, "The Origin of the Family, Private Property, and the State," in Marx and Engels, *Selected Works*, 2 vols (Moscow, 1958), v. II, p. 324. I do not believe that the workers had been better off as slaves, as does Gilberto Freyre. Certainly the fact that Pernambuco's free colored population succeeded in reproducing itself and increasing in numbers indicates that freedom was preferable to slavery. But there were very few differences in working conditions. In this respect, I agree with the conclusions of A. J. R. Russell-Wood, "Colonial Brazil," in David W. Cohen and Jack P. Green, eds., *Neither Slave nor Free* (Baltimore, 1972), p. 132.

31. José Lins do Rego, *Banguê* (Rio de Janeiro, 1934).

All forms of free employment entailed the basic liberty to leave the job, a right denied to slaves; but this freedom should not be exaggerated. The free worker did not find alternative employments in the northeast, as did ex-slaves in Caribbean plantation societies after abolition.[32] Census data do not indicate any redistribution of the free colored population toward Pernambuco's west, where they might have entered subsistence crop farming because of the greater availability of land and the smaller population; nor did workers move out of the rural areas into the capital, Recife. Census data do show a much faster rate of native population growth in Brazil's center-south provinces than in the northeast. This growth may reflect superior economic conditions making for higher natural increase, or it may indicate the migration south of northeastern workers, and thereby a much higher degree of labor mobility than is presently thought. The source of this growth awaits further study; in any event, most ex-slaves stayed in Pernambuco, and in the sugar zone.[33]

The lack of jobs and the unattractiveness of free labor conditions on plantations led to constant vagabondage. The vagabonds took jobs only infrequently, and the rest of the time they lived off the land. Since the sugar planters virtually monopolized the land, in effect the vagabonds lived by stealing. Even when ex-slaves took jobs, their previous work experience had left a bitter taste and predisposed them to value leisure highly. Hence their supply curve bent backward, and they "bought" leisure by sacrificing steady employment. Even in the center-south, where general economic conditions far surpassed those prevailing in Pernambuco, vagabondage occurred. The superabundant work force allowed sugar planters to reject many workers; but the poor working conditions also led many workers to reject steady employment on the plantations.[34]

32. Gwendolyn Midlo Hall, *Social Control in Slave Plantation Societies: A Comparison of St. Domingue and Cuba* (Baltimore, 1971), pp. 120-121. J. H. Parry and P. M. Sherlock, *A Short History of the West Indies*, 3rd edition (London, 1971, first published in 1956), pp. 195-196.

33. Nathaniel Leff argues transport costs prevented such labor mobility. "Economic Development and Regional Inequality," p. 254.

34. Toplin, *The Abolition of Slavery in Brazil*, pp. 259-261, suggests the scope of post-abolition vagabondage.

Some Pernambucans thought to improve the quality of rural labor by encouraging European immigration; but European immigrants found little reason to settle in Pernambuco. The hot climate in the sugar zone, the scarcity of land, and principally the unfavorable economic conditions in the sugar industry all discouraged immigrants. Slavery and discriminatory civil legislation did not particularly keep Europeans out of Pernambuco any more than São Paulo. But the booming coffee economy permitted subsidizing hundreds of thousands of Europeans and these newcomers took the chance of settling there because they could hope to improve their lots in the growing economy. The immigrants began working in coffee, but they soon moved into other occupations, accumulated capital, and frequently played important roles in the center-south's industrialization. The absence of such an economically active group in Pernambuco may have debilitated that region, but we should not forget that the immigrants' activity was probably more a function of the vigorous coffee economy than of inherent talents.

Just as a different government might have cushioned or avoided the export market crisis, so too a different government might have acted more positively to ease the transition from slave to free labor. The government saw fit to offer compensation through emancipation funds to slave owners before 1888. While it refused to indemnify for final abolition, the government made no attempt to aid the ex-slaves. In the United States, a coalition of moderate and radical reconstructionists in Congress overrode President Andrew Johnson's veto and created a Freedmen's Bureau. Radicals advocated the distribution of vacant and war-confiscated lands, which led to the slogan "40 acres and a mule," but the bureau lasted only a year, and little land reform occurred.[35] In Brazil, abolition did not occur amidst the destruction of a civil war, and the Brazilian government could completely ignore the ex-slaves after 1888. This neglect left the freedmen on northeastern plantations as before; in the center-south, it meant

35. Kenneth M. Stampp, *The Era of Reconstruction, 1865-1877* (New York, 1965), pp. 112, 125-135.

234   THE SUGAR INDUSTRY IN PERNAMBUCO

that as European immigrants won the better jobs, freedmen remained at the bottom of society.[36]

Even a successful Freedmen's Bureau in Brazil, however, would not have guaranteed the improvement of the workers' welfare without massive government support. An even more liberal reform, backing worker organization for greater bargaining power, might have yielded short-run benefits, but in the end plantation unions would only have divided the shrinking sugar revenues more evenly, without making long-term provision for the general welfare. Moreover, they would have been constantly subject to the planters' influence in local government.[37]

A thorough-going land reform, which not only changed crops but also redistributed land, might have yielded a more equitable distribution of income and raised workers' welfare; but it was not politically feasible. A government less subject to planter control might have turned to land reform combined with diversification to avoid the income effects of sugar's stagnation. A Pernambucan growing a wider variety of export crops, such as sugar, cotton, and tobacco, as well as food for regional consumption, would have remained relatively immune to violent changes in any one commodity market while at the same time he might have met importers' needs for foreign exchange. More marginal lands could have been used for livestock, poultry, or vegetables, improving the general nutritional level. The government would have supplied the substantial credit entailed in such a land reform in the form of

36. For an introduction to the plight of the Brazilian Negro after abolition, particularly in São Paulo, see Florestan Fernandes, *The Negro in Brazilian Society* (New York, 1969, first published 1965), pp. 1-20. For an opposite view, see Herbert Klein, "Nineteenth Century Brazil," in Cohen and Greene, *Neither Slave nor Free*, p. 332.

37. On the tentative efforts of peasant leagues to fight the planters in the 1960's, see Leda Barreto, *Julião–Nordeste–Revolução* (Rio de Janeiro, 1963), and Cynthia N. Hewitt, "The Peasant Movement of Pernambuco, Brazil: 1961-64," in Henry A. Landsberger, ed., *Latin American Peasant Movements* (Ithaca, N.Y. 1969), pp. 374-398. In 1964 the coup d'etat by a military group hostile to labor mobilization crushed the peasant leagues. For a good bibliography of recent Brazilian peasant movements, see the notes in Shepard Forman, "Disunity and Discontent: A Study of Peasant Political Movements in Brazil," in Chilcote, *Protest and Resistance in Angola and Brazil*, pp. 183-205.

many small loans, instead of the few large subsidies granted to the usineiros, which after all were largely defaulted upon anyway. The land required would have been taken from the planters, thereby increasing either the number of landholdings without creating minifundia, or else the number of landholders through cooperative or collective arrangements. The result would have been a reduction in gross regional product, but a more equitable distribution of that product. Such reform in nineteenth-century Brazil of course was politically unrealistic, although not unheard of, as the writings of Antônio Pedro de Figueiredo in 1847 and 1848 testify.

If this study emphasizes one theme, it is that modernization involving capital improvements and reorganization, and the abolition of forced labor, did not always bring either economic or social change. Brazil's northeastern sugar regions made certain necessary adjustments in the last quarter of the nineteenth century, but these only served to soften the impact of unfavorable foreign markets on the planter class and preserve the traditional economic and social structure. Where other planter groups enjoyed a more successful experience in this period, as in Cuba and São Paulo, the principal factor governing their successes was the strength of their export markets.

The only alternatives to stagnation for Pernambuco entailed recolonization or radical structural reform; in both cases the costs would have been borne by the planters. With a land reform, in the short run the region's product might have suffered as well, but in the long run the majority of the population might have benefited. The planters remained dominant politically, and followed their clear economic self-interest. But the market, in conjunction with unavoidable political distortions, precluded a socially efficient pattern of development.

Brazil is hardly the only colonial or semicolonial area in which modernization has preserved the traditional structure. It was no accident that Cuba, one of the most advanced sugar colonies in the Western Hemisphere, both in terms of technology and the development of labor organizations, was also the first to experience a successful socialist revolution. No amount of capital investment or revision of labor modes in capitalist agricultural export economies can yield social benefits unless the changes increase the workers' control and shares in production. Cuba began a revolution in 1959. Pernambuco went through several adjustments in the last quarter of the nineteenth century, but it still awaits a more generally beneficial modernization.[38]

38. For recent testimony that the situation described for the late nineteenth century has not basically changed, see Marvine Howe, "Brazil Peasants Find Their Plight Worsens," *New York Times*, November 27, 1972, p. 12.

APPENDIX 1

WORLD BEET SUGAR PRODUCTION

(in metric tons)

Years	France	Germany	Austria-Hungary	Russia	Holland
1841-45	33,291	13,181			
1846-50	59,015	35,708			
1851-55	71,535	76,988		17,720	
1856-60	118,929	128,151	74,000	18,440	
1861-65	168,612	154,268	106,400	50,400	
1866-70	239,998	210,553	162,671	108,000	
1871-75	400,584	266,795	273,726	213,125	26,098
1876-80	320,306	420,130	403,464	301,960	24,797 [a]
1881-85	350,583	894,451	437,600	331,000	
1886-90	501,183	1,111,862	643,493	451,200	
1891-95	584,491	1,460,412	845,434	542,400	72,600
1896-1900	809,082	1,838,806	994,874	751,700	139,298
1901-05	820,704	2,002,415	1,161,681	1,059,601	139,687
1906-10	689,312	2,218,316	1,378,197	1,469,062	179,793

(Continued)

Years	Belgium	Italy	Denmark	Sweden	U.S.
1841-45					
1846-50					
1851-55					
1856-60					
1861-65					
1866-70					
1871-75	96,823				
1876-80	61,879 [a]				
1881-85	79,694		8,756		
1886-90	122,531		19,636	11,824	1,641
1891-95	196,015	1,706	30,231	44,266	17,700
1896-1900	263,600	20,453	43,688	96,470	51,960
1901-05	252,819	104,752	51,068	102,207	212,720
1906-10	261,564	151,685	69,514	142,154	427,600

[a] 1880 not available.

SOURCES: Deerr, *The History of Sugar*, v. II, pp. 490-498. Hermann Paasche, *Zuckerindustrie und Zuckerhandel der Welt* (Jena, 1891) p. 347.

APPENDIX 2

WORLD CANE SUGAR PRODUCTION

(in metric tons)

Years	Brazil [a]	Cuba	Puerto Rico	British Guiana	Martinique, Guadeloupe	U.S. (Louisiana)	Mexico
1841-45	89,188			32,537	60,260	73,939	
1846-50	118,287			33,478	44,984	109,276	
1851-55	123,409	320,722		44,455	43,467	162,039	
1856-60	106,243	394,200		50,125	55,040	120,150	
1861-65	126,763	534,600		65,449	53,607	67,151	
1866-70	105,939	682,000		74,593	65,537	39,927	
1871-75	170,543	682,200		84,287	80,335	61,108	
1876-80	175,599	568,600		93,010	84,957	95,212	
1881-85	228,302	505,215		110,884	94,559	113,310	
1886-90	155,993	621,696		113,154	82,267	147,957	
1891-95	153,333	933,470	50,625 [b]	107,586	68,390	243,789	
1896-1900	113,908	272,427	52,198	96,828	70,950	256,266	73,367 [c]
1901-05	78,284	943,212	136,957	116,741	68,705	319,466	109,140
1906-10	51,338	1,393,898	255,829	110,809	76,247	316,283	139,400

(Continued)

239

Years	Peru	Argentina	Mauritius	Java	Queensland	Philippines	Hawaii
1841-45			35,238	61,570			
1846-50			59,258	90,392			
1851-55			88,435	104,827		37,720	
1856-60			122,157	126,783		41,400	
1861-65			122,155	138,370		48,660	3,294
1866-70			102,339	157,816		70,584	8,077
1871-75	33,333 [d]		106,739	198,873		98,420	9,976
1876-80	78,820 [e]		119,577	231,571	13,640	137,158	18,058
1881-85	23,050		115,113	321,469	31,627	180,615	57,160
1886-90	49,140		121,267	359,056	52,647	180,430	104,100
1891-95	58,740	76,213	115,705	484,198	67,342	255,800	130,200
1896-1900	98,440	111,526	158,582	670,485	113,853	128,639	228,000
1901-05	137,447	111,947 [f]	169,275	947,975	117,221	136,167	345,600
1906-10	140,502	131,676	204,899	1,209,098	171,561	372,726	436,400

[a] Exports only.
[b] 1891 not available.
[c] 1896-97 not available.
[d] 1871-72 not available.
[e] 1880-81 not available.
[f] 1904 not available.

SOURCES: "O Açúcar na vida econômica do Brasil," pp. 233-235. H. C. Prinsen Geerligs, *The World's Cane Sugar Industry. Past and Present* (Manchester, 1912), pp. 120-121, 156-157, 176, 262, 242. Paasche, *Zuckerindustrie und Zuckerhandel der Welt*, pp. 348-349. Deerr, *The History of Sugar*, v. I, pp. 131, 143, Cuban Economic Research Project, *A Study on Cuba*, p. 83.

APPENDIX 3

PERNAMBUCO SUGAR MILLS

Region	Years								
	1761-75	1844	1850's	1860's	1870's	1880's	1890's	1900's	1914
1. *Tablelands*									
Recife	10								4
Afogados		13	7	10	11	9		7	
Poço da Panella		3	1	1	1				
Várzea			11	11	15	12		8	
Olinda	15	3		6	1		10	6	9
Maranguape	4	6	6		7	3			
Sé						2			
Goiana	24	55	73	91	32	35	86	93	96
N.S. do O				34	37	45	50		
N.S. do Rosário					34				
São Lourenço de Tejucupapo			12	28	28	9	10	9	
Igarassú	16	34	46	47	47	52	56		52
Itamaracá	9	3	5	5	5	6	10	6	3
Itambé	5		21	24	24	45	50	47	58
Total	83	117	182	257	242	218	272	176	222

(*Continued*)

	1761-75	1844	1850's	1860's	1870's	1880's	1890's	1900's	1914
2. Dry Mata									
Glória do Goitá			15	19	17	19	60	19	16
Nazaré		68	89	114	85	87	256	60	199
Pau d'Alho		39	38	53	43	45	61		50
São Lourenço da Mata	20	28	44	34	28	22	56	59	51
N.S. da Luz			27	30	31	32		33	
Timbauba						27	70	49	75
S. Vicente				40	33	36		33	
Tracunhaém	17		80	97	84	101		56	
Total	54	135	293	387	321	369	523	309	391
3. Humid Mata									
Água Preta		44	63	97	104	129	140	139	140
Amarají							66	65	60
Barreiros			46	51	53	55	50	53	47
Bonito		3	24	45		96		93	63
Cabo	26	86	62	80	79	79	95	78	73
Escada			103	108	109	120	120	80	84
Gameleira					56	57	67	69	71
Ipojuca	30		54	63	65	66	62		61
Jaboatão	14	34	46	48	48	52	71	71	61
Muribeca	8		18	20	18	20	22		
Palmares					113	99	100	134	112

Rio Formoso	37		42	42	36	36	68	77	73
Sirinhaém		26	87	96	73	69	64	56	62
Vitória de Santo Antão	4	85	66	82	45	81	90	78	66
Total	119	344	623	732	799	959	1015	993	973
4. Agreste									
Bezerros			13	8				36	12
Brejo			6	3	4				40
Alagoa de Baixo									5
Cimbres			12	13					4
Caruarú			47						9
Altinho			29						6
Panellas			12						85
Quipapá			24	23					88
Garanhuns		1	16						61
Correntes									142
Ilha das Flores			6	6					
Limoeiro		8	26	12	13	13	15		14
Bom Jardim	1		27	42	30	38	55		51
Taquaratinga				36	5	16	15	16	3
Papacaça			22	31					
Taquara	7			34					
Una	23	65	30	41	32	41	80		36
Total	31	74	258	249	84	108	165	52	556

(*Continued*)

5. Sertão

	1761-75	1844	1850's	1860's	1870's	1880's	1890's	1900's	1914
Águas Belas									19
Belmonte									9
Boa Vista									32
Buique									36
Cabrobó									27
Exú									24
Flores				5					76
Floresta									4
Granito									10
Afogados da Ingazeira									65
Leopoldina									8
Pedra									1
Petrolina									12
Salgueiro									54
Tacaratú									150
Espiritu Santo				42					
Triunfo									88
Vila Bela									31
Total				47					646
Grand Total	296	670	1,356	1,672	1,446	1,654	1,975	1,530	2,788

NOTE: For qualifications of these data, see notes for Table 21. Counties listed are those existing in 1900. Parishes and suburbs are subsumed under the respective county—for example, Afogados under Recife. Sudden appearances of many engenhos, such as in Amarají and Palmares, reflect the creation of separate counties, not engenho-building sprees.

SOURCES: 1761: Letter from Luiz Diogo Lobo da Silva, Recife, February 15, 1761, in Correspondência do Governador de Pernambuco, 1753-1770, Instituto Historico e Geographico do Brasil, Lata 1-1-14, fólios 249-260. I am indebted to Herbert Klein for a copy of this document.

1775: "Idéa da População da Capitania de Pernambuco, e das suas annexas, extensão de suas costas, Rios e Povoações notáveis. Agricultura, numero dos Engenhos, contractos, e Rendimentos Reaes, augmento que estes tem tido etc. etc. desde o anno de 1774 em que tomou posse do Governo das mesmas Capitanias o Governador e Capitam Geral General Jozé Cesar de Menezes," Annaes de Bibliotheca Nacional do Rio de Janeiro, v. 40 (1918), pp. 1-111.

1844: Figueira de Mello, Ensaio sôbre a Statistica, p. 263. 1850's, 1860's, 1870's, 1880's, 1890's, 1900's: José de Vasconcellos, Almanack . . . 1860, 1861, 1862. F. P. do Amaral, Almanack . . . 1868, 1869, 1870, 1872, 1875, 1876, 1881, 1884, 1885, 1886. Verissimo de Toledo, Almanack . . . 1893, 1894, 1895. Pires Ferreira, Almanack . . . 1899, 1900, 1901. For 1914: Peres and Peres, A Industria Assucareira em Pernambuco, pp. 32-33.

Instituto Brasileiro de Geografia e Estatística, Encyclopedia dos Municipios Brasileiros, 36 vols. (Rio de Janeiro, 1957-58), v. XVIII, Pernambuco. Sebastião de Vasconcellos Galvão, Diccionario Chorographico, Histórico e Estatistico de Pernambuco, 4 vols. (Rio de Janeiro, 1908-1927).

GLOSSARY

$\mathcal{T}$he orthography of Brazilian Portuguese is everchanging. We have preserved the original in cases of direct citations and proper names of people and organizations who disappeared before the orthography changed. In all other instances we have used the orthography standard in 1971.

agreste	geographical region west of the sugar zone
barcaça	sailing vessel employed in coastal trade
barão	baron
branco	white
caatinga	agreste
camumbembe	squatter
cangaceiro	rural outlaw
cangaço	cangaceiro's life style
capanga	henchman
centrales (Spanish)	central mills
chã	elevated plateau
colono (Sp.)	Cuban cane supplier
comissário	factor
conto	unit of Brazilian currency

correspondente	comissário
demerara	type of yellow sugar
encilhamento	period of financial speculation in the early 1890's
engenho	traditional mill
fazendeiro	owner of a large agricultural property
feixe	bundle of cut cane
fogo morto	plantation whose mill no longer operates
folheto	pamphlet of popular poetry
fornecedor	supplier of sugar cane
ingenio (Sp.)	traditional mill
jagunço	capanga
jangada	sailing raft
lavrador	sharecropper
literatura de cordel	folheto
mascavado	type of crude sugar
massapê	soils good for sugar cane growing
mil-réis	unit of Brazilian currency
moléstia	disease
morador de condição	squatter working for wages
parceiro	smaller sharecropper
pau brasil	Brazil wood, used for dyes
poder moderador	Brazilian Emperor's power to appoint and dismiss ministers and dissolve the imperial legislature
Praieiro	participant in the Pernambuco revolution of 1848-49
propriedade	farm
quinguingu	time when the slave worked for himself
rendeiro	larger sharecropper
senhor de engenho	owner of an engenho. We use "planter" to refer generally to this type, and "mill owner" to refer specifically to sugar manufacturers.
sertanejo	native of the sertão

sertão	geographical region west of the agreste, prone to drought
sesmaria	land grant
sítio	farm
usina	modern sugar factory
usineiro	usina owner
várzea	fertile river valley lands
visconde	viscount
zona da mata	principal sugar-growing region

MEASURES

WEIGHTS

1 *arroba*	15	kilograms
1 loaf (sugar)	63.4	kilograms
1 sack (sugar)	75	kilograms
1 sack (cotton)	85	kilograms
1 barrel (sugar)	120	kilograms
1 chest (sugar)	300	kilograms
1 ton ("long ton")	1,000	kilograms

VOLUMES

1 *cuia*	1.1	liters
1 *canada*	2.77	liters
1 *alqueire*	36.4	liters
1 *pipa*	485	liters

DISTANCES

1 *palmo*	0.22	meters
1 *braça*	2.2	meters
1 *légua de sesmaria*	6.6	kilometers

AREAS

1 *square braça*	4.84	square meters
1 hectare	10,000	square meters
1 *légua*	43.56	square kilometers

CURRENCIES

1 *mil-réis* (1$000)   1,000 *réis*
1 *conto* (1:000$000)   1,000 mil-réis
1 English pound (£)   20 shillings of 12 pence each

NOTES: Brazilian measures in the nineteenth century varied widely from province to province, and even within provinces between regions. This table gives those measures most commonly found in available Pernambuco sources. All equivalents are in the metric system officially adopted in Brazil in 1874.

All currency equivalents are given in English money, which was the standard foreign currency in the nineteenth century. For U.S. dollar equivalents of Brazilian currency, see Julian Smith Duncan, *Public and Private Operation of Railways in Brazil* (New York, 1932), pp. 183-184. For a rough idea of the dollar exchange rate, a mil-réis was worth $.55 at the official par (27d/1$000), and a conto was worth $550.

SOURCES: Jerônymo Martiniano Figueira de Mello, *Ensaio sôbre a Statística Civil e Política da Província de Pernambuco* (Recife, 1852). *Relatório da Direcção da Associação Commercial Beneficente de Pernambuco apresentado à Assembléa Geral da mesma em 5 de agôsto de 1873* (Recife, 1873). *O Imperio do Brazil na Exposiçao Universal de 1873 em Vienna d'Austria* (Rio de Janeiro, 1873), pp. 170-175. Sociedade Auxiliadora da Agricultura de Pernambuco, *Boletim,* no. 3 (September 1882), pp. 53-54. Verissimo de Toledo, *Almanack . . . 1893,* pp. 49-54. Ministerio da Agricultura, Industria e Commercio, Servicos de Inspecção e Defesa Agricolas. *Questionarios sobre as condições da agricultura dos 173 municipios do Estado de Sao Paulo. De abril de 1910 a janeiro de 1912* (Rio de Janeiro, 1913), pp. 549-550.

BIBLIOGRAPHY

𝒯his bibliography includes only sources cited in the text, not all sources consulted.

Archives

Associação Comercial Beneficente de Pernambuco, Recife.
Bancroft Library, University of California, Berkeley. Photograph Collection.
Cartório Público de Ipojuca, Pernambuco.
Instituto Histórico e Geográfico do Brasil, Rio de Janeiro. Correspondência do Governador de Pernambuco, 1753-1770. Lata 1-1-14, folios 249-260.
————. Photograph Collection.
Museu do Açúcar, Recife. Photographs in Coleção Francisco Rodrigues.
Pernambuco, Arquivo Público Estadual, Recife. Coleção Camaras Municipaes.
————. Coleção Engenhos Centraes.
————. Coleção Ministério da Agricultura.
————. Coleção Obras Públicas.
————. Coleção Portarias.
————. Coleção Tesouro do Estado.
————. Registro de Legitimação de Posse de Terras.
————. Registro de Terras Públicas.
Sociedade Auxiliadora da Agricultura de Pernambuco, Recife. Livro de Atas no. 1 (Assembléa Geral).
————. Livro de Atas no. 2 (Conselho Administrativo).

U.S. National Archives, Washington, D.C. Despatches from United States Consuls in Pernambuco, 1817-1906. Microcopy T344.

Usina União e Indústria, Recife and Escada.

Sterling Memorial Library, Yale University. New Haven, Connecticut. Map Collection.

Unpublished Materials

Dantas, Bento. "A Agro-Indústria Canavieira de Pernambuco: As Raizes Históricas dos Seus Problemas, Sua Situação Atual e Suas Perspectivas." Mimeographed, 1968.

Denslow, David. "Slave Mortality." Mimeographed, 1969.

Eisenberg, Peter L. "The Sugar Industry of Pernambuco, 1850-1889." Ph.D. dissertation at Columbia University, New York, 1969.

_____, and Michael M. Hall. "Labor Supply and Immigration: A Comparison of Pernambuco and São Paulo." Paper presented to the 4th Annual Meetings of the Latin American Studies Association, Madison, Wisconsin, May 1973.

Graham, Douglas H. and Sergio Buarque de Hollanda Filho. "Migration, Regional and Urban Growth and Development in Brazil: A Selective Analysis of the Historical Record—1872-1970." Mimeographed, 1971.

Guerra, Flávio. "Memórias de uma Associação (História do Comércio do Recife)." Unpublished manuscript, 1965.

Hall, Michael M. "The Italians in São Paulo, 1880-1920." Paper presented to the American Historical Association annual meetings, New York,, December 1971.

_____. "The Origins of Mass Immigration in Brasil, 1871-1914." Ph.D. dissertation at Columbia University, New York, 1969.

Leff, Nathaniel H. "Economic Retardation in Nineteenth Century Brazil." Mimeographed, 1970.

Perucci, Gadiel. "Le Pernambouc (1889-1930): Contribution à l'Histoire Quantitative du Brésil." Thèse de doctorat de 3ème cycle. Paris, 1972.

Prebisch, Raul. Interview, New Brunswick, New Jersey, May 20, 1971.

Brazilian Newspapers and Magazines

Anuário Açucareiro. Rio de Janeiro.
O Agricultor Prático. Recife.
Brasil Açucareiro. Rio de Janeiro.

O Brazil Agricola, Industrial, Commercial, Scientifico, Litterario, e Noticioso. Recife.

Diário de Pernambuco. Recife.

A Cidade. Recife.

Correio de Recife.

Commercio de Pernambuco. Recife.

Economia e Agricultura. Rio de Janeiro. Later titled *Brasil Açucareiro.*

A Escada.

O Escadense Periódico Político. Escada.

O Industrial: Revista de Indústrias e Artes. Recife.

Jornal do Recife.

Pernambuco. Recife.

O Progresso. Recife.

A Província. Recife.

Revista do Instituto Arqueológico, Histórico e Geográfico Pernambucano. Recife.

Sociedade Auxiliadora de Agricultura de Pernambuco, *Boletim.* Recife.

Published Materials

"O Açúcar na vida econômica do Brasil," *Annuário Açucareiro para 1938.* pp. 7-288.

Additamento às Informações sôbre o Estado de Lavoura. Rio de Janeiro, 1874.

Agarwala, A. N., and S. P. Singh (editors), *The Economics of Underdevelopment.* New York, 1963.

Alden, Dauril, "The Population of Brazil in the late Eighteenth Century: A Preliminary Survey," *Hispanic American Historical Review,* (Durham, North Carolina) v. XLIII, no. 2 (May 1963), pp. 173-205.

Alves Martins, Francisco. "Antonio Silvino e o Negro Currupião." n.p., n.d.

Amaral, F. P. do. *Almanack Administrativo, Mercantil, Industrial e Agrícola da Província de Pernambuco para o anno de 1868, 1869, 1872, 1873, 1875, 1876, 1881, 1884, 1885, 1886.* Recife, 1868-86.

Amaral, Luis, *História Geral da Agricultura Brasileira no Triplice Aspecto Político Socio-Econômico,* 2nd edition, 2 volumes. São Paulo, 1958. First published 1940-41.

Andrews, Christopher Columbus. *Brazil, Its Conditions.* New York, 1887.

Annaes da Assembléa Provincial de Pernambuco. Recife, 1862-89.

Annaes do Senado do Estado de Pernambuco. Recife, 1896-98.

Annuário Commercial Pernambuco, Parahyba, Alagôas, Bahia, 1902-03. Recife.

Arriaga, Eduardo E. *New Life Tables for Latin American Populations in the Nineteenth and Twentieth Centuries*. Berkeley, 1968.

Arruda Beltrão, Antonio Carlos de. *A Lavoura da Canna e a Indústria Assucareira Nacional*. Rio de Janeiro, 1918.

Associação Comercial Beneficente de Pernambuco. *Relatório da Direcção da Associação Commercial Beneficente de Pernambuco apresentado à Assembléa Geral da mesma*. Recife, 1857-1909.

Auler Guilherme, *Os Utinga, Filhos, Netos, e Bisnetos do Senhor do Engenho Matapiruma*. Recife, 1963.

Azevedo, Fernando de. *Canaviais e Engenhos na Vida Política do Brasil*. Rio de Janeiro, 1948.

Baer, Werner. *Siderugia e Desenvolvimento Brasileira*. Translated by Wando Pereira Borges. Rio de Janeiro, 1970. First published as *The Development of the Brazilian Steel Industry*, 1970.

Banco de Crédito Real de Pernambuco. *Relatório do Banco de Crédito Real de Pernambuco apresentado à Assembléa Geral dos accionistas em 23 de março de 1889*. Recife, 1889.

———. *Relatório apresentado à Assembléa Geral dos Accionistas em 29 de maio de 1912*. Recife, 1912.

Baptista, Francisco das Chagas. "A política de Antonio Silvino." Recife, 1908.

Barbosa Lima, Alexandre José. *Discursos Parlamentares*, 2 volumes. Brasília, 1963-68.

———. *Mensagens apresentadas ao Congresso Legislativo do Estado em 1893, 1895, e 1896 pelo Dr. Alexandre José Barbosa Lima, quando Governador de Pernambuco*. Recife, 1931.

Barbosa Lima Sobrinho, Alexandre José. "O Govêrno Barbosa Lima e a Indústria Açucareira de Pernambuco," *Annuário Açucareiro para 1938*. pp. 353-367.

———. *Problemas Econômicos e Sociais da Lavoura Canavieira*, 2nd edition. Rio de Janeiro, 1943.

———. *A Revolução Praieira*. Recife, 1949.

Barrata Góes, Manoel. *Nucleo Colonial Suassuna, O Delegado da Inspectoria Geral das Terras e Colonisação ao Excm. Sr. Governador do Estado e ao Público*. Recife, 1894.

Bareto, Leda. *Julião–Nordeste–Revolução*. Rio de Janeiro, 1963.

Bello, José Maria. *Memórias*. Rio de Janeiro, 1958.

Bello, Júlio. *Memórias de um Senhor de Engenho*. Rio de Janeiro, 1938.

Beiguelman, Paula. "O encaminhamento político do problema da escravidão no império," in Buarque de Holanda, *História Geral da Civilização Brasileira,* tomo II, v. 3, pp. 189-201.

Benévolo, Ademar. *Introdução à História Ferroviária do Brasil. Estudo social, político e histórico.* Recife, 1953.

Bennett, Frank. *Forty Years in Brazil.* London, 1914.

Bethell, Leslie. *The Abolition of the Brazilian Slave Trade. Britain, Brazil and the Slave Trade Question, 1807-1869.* Cambridge, England, 1970.

Boxer, C. R. *The Golden Age of Brazil, 1695-1750: Growing Pains of a Colonial Society.* Berkeley, 1961.

Brandão Sobrinho, Júlio. *A Lavoura da Canna e a Indústria Assucareira dos Estados Paulista e Fluminense. Campos e Macahé em confronto com S. Paulo. Relatório apresentado ao Illm. e Exm. Sr. Dr. Antônio de Pádua Salles.* São Paulo, 1912.

Brazil. Congresso. Camara dos Deputados. *Reforma Hypothecaria.* Rio de Janeiro, 1856.

Buarque de Holanda, Sérgio (editor). *História Geral da Civilização Brasileira.* 6 volumes, São Paulo, 1963-71. Tomo I, *A Época Colonial,* v. 2, *Administração, Economia, Sociedade.* Tomo II, *O Brasil Monárquico.* v. 1, *O Processo de Emancipação;* v. 2, *Dispersão e Unidade;* v. 3, *Reações e Transações;* v. 4, co-edited by Pedro Moacyr Campos, *Declínio e Queda do Império.*

Buescu, Mircea. *História Econômica do Brasil, Pesquisas e Análises.* Rio de Janeiro, 1970.

————. and Vicente Tapajós. *História do Desenvolvimento Econômico do Brasil.* Rio de Janeiro, n.d.

Burke, Ulick Ralph, and Robert Staples, Jr. *Business and Pleasure in Brazil.* London, 1884.

Burlamaque, F. L. C. *Monographia da Canna d'Assucar.* Rio de Janeiro, 1862.

Burns, E. Bradford. (editor). *A Documentary History of Brazil.* New York, 1966.

————. *A History of Brazil.* New York, 1971.

————. *The Unwritten Alliance. Rio Branco and Brazilian-American Relations.* New York and London, 1966.

Caldas Lins, Rachael, and Gilberto Osório de Andrade. *As Grandes Divisões da Zona da Mata Pernambucana.* Recife, 1964.

Calógeras, João Pandiá. *A History of Brazil.* Translated and edited by Percy Alvin Martin. Chapel Hill, North Carolina, 1939.

————. *A Política Monetária do Brasil*. Translated by Thomaz Newlands Neto. São Paulo, 1960.

Camilo, Manoel. "O Grande e Verdadeiro Romance de Antonio Silvino." Campina Grande, Paraíba, n.d.

Canabrava, Alice P. "A Grande Lavoura," in Buarque de Holanda and Moacyr Campos, *História Geral da Civilização Brasileira*, tomo II, v. 4, pp. 85-137.

————. "A Grande Propriedade Rural," in Buarque de Holanda, *História Geral da Civilização Brasileira*, tomo I, v. 2, pp. 192-217.

Cardoso, Vicente. *À margem da história do Brasil*. São Paulo, 1933.

Carneiro, Edison. *A Insurreição Praieira, 1848-1849*. Rio de Janeiro, 1960.

Carneiro Rodrigues Campello, Samuel. *Escada e Jaboatão: Memória apresentada ao VI Congresso de Geographia Brasileiro*, Recife, 1919.

Carneiro Vilela, Joaquim Maria. "O Clube do Cupim," *Revista do Instituto Arqueológico, Histórico e Geographico Pernambucano*. v. XXVII, nos. 127-130 (1925-26), pp. 417-427.

Castro, Josué de. *Documentário do Nordeste*. 3rd edition, São Paulo, 1965. First published 1937.

————. *Geografia da Fome (O Dilema Brasileiro: Pão ou Aço)*. 9th ed., São Paulo, 1965. First published 1946.

Catálogo da Exposição realizada no Teatro Santa Izabel de 13 à de Maio de 1938. Recife, 1939.

Cavalcanti, Amaro. "Antecedentes Históricos do Estatuto da Lavoura Canavieira," *Jurídica* (Rio de Janeiro), v. xxvi, no. 70 (October-December 1962), pp. 324-332.

Colleção de Leis Estaduais de Pernambuco. Recife, 1890-95.

Colleção de Leis do Império do Brazil. Rio de Janeiro, 1875-89.

Colleção de Leis Provinciais de Pernambuco. Recife, 1875-89.

Commissões da Fazenda e Especial. *Parecer e Projecto sôbre a Creação de Bancos de Crédito Territorial e Fábricas Centraes de Assucar apresentados à Câmara dos Srs. Deputados na Sessão de 20 de Julho de 1875*, Rio de Janeiro, 1875.

Conrad, Robert. "The Contraband Slave Trade to Brazil, 1831-1845," *Hispanic American Historical Review*, v. XLIX, no. 4 (November 1969), pp. 617-638.

————. *The Destruction of Brazilian Slavery, 1850-1888*. Berkeley and Los Angeles, 1972.

Conrad, Alfred H., and John R. Meyer. *The Economics of Slavery and other Studies in Econometric History*. Chicago, 1964.

BIBLIOGRAPHY 257

Corbitt, Duvon Clough. *A Study of the Chinese in Cuba, 1847-1947*. Wilmore, Kentucky, 1971.

Correia de Andrade, Manuel. *Paisagens e Problemas do Brasil*. São Paulo, 1968.

———. *A Terra e o Homem do Nordeste*. 2nd edition, São Paulo, 1964. First published 1963.

Correia Lopes, Edmundo. *A Escravatura (subsidios para a sua história)*. Lisbon, 1944.

Costa Filho, Miguel. *A Cana-de-Açucar em Minas Gerais*. Rio de Janerio, 1963.

———. "Engenhos Centrais e Usinas," *Revista do Livro* (Rio de Janerio), Ano V., no. 19 (September 1960), pp. 83-91.

Costa Porto, *Estudo sôbre o Sistema Semarial*. Recife, 1965.

Cruz Gouveia, Fernando da. "Os De Mornay e a Indústria Açucareira em Pernambuco," *Brasil Açucareiro*, ano XXXV, v. LXX (August 1967), pp. 78-84.

Cuban Economic Research Project. *A Study on Cuba*. Miami, 1965.

Curtin, Philip D. *The Atlantic Slave Trade. A Census*. Madison, Wisconsin, 1969.

Davis, David Brion. *The Problem of Slavery in Western Culture*. Ithaca, New York, 1966.

Dé Carli, Gileno. "Alagôas. Sinópse Histórica do Açucar," *Annuário Açucareiro para 1935*. pp. 36-39.

———. *Aspectos Açucareiros de Pernambuco*. Recife, 1940.

———. *Aspectos da Economia Açucareira*, Rio de Janeiro, 1942.

———. *Evolução do problema canavieiro fluminense*. Rio de Janeiro, 1942.

———. *Gênese e evolução da industria açucareira paulista*. Rio de Janeiro, 1943.

———. "Geografia Económica e Social da Canna de Açúcar no Brasil," *Brasil Açucareiro*, v. X, no. 2 (October 1937) pp. 200-226.

———. *O Processo Histórico da Usina em Pernambuco*. Rio de Janeiro, 1942.

Dean, Warren. *The Industrialization of São Paulo, 1880-1945*. Austin, Texas, 1969.

———. "Latifundia and Land Policy in Nineteenth-Century Brazil," *Hispanic American Historical Review*, v. 51, no. 4 (November 1971), pp. 606-625.

Decretos do Governo Provisório da República dos Estados Unidos do Brazil. Rio de Janeiro, 1890-91.

Deerr, Noel. *Cane Sugar: A Textbook on the Agriculture of the Sugar Cane, the Manufacture of Cane Sugar, and the Analysis of Sugar House Products.* 2nd edition, London, 1921.

———. *The History of Sugar.* 2 volumes, London, 1949-50.

Degler, Carl N. *Neither Black nor White, Slavery and Race Relations in Brazil and the United States.* New York, 1971.

———. "Slavery in Brazil and the United States: an Essay in Comparative History," *American Historical Review* (Washington, D.C.), v. LXXV, no. 4 (April 1970), pp. 1,004-028.

Della Cava, Ralph. *Miracle at Joazeiro.* New York, 1970.

Denis, Pierre. *Brazil.* Translated by Bernard Miall. London, 1911.

Dent, Hastings Charles. *A Year in Brazil.* London, 1886.

Diégues Júnior, Manuel. *O Bangüé nas Alagôas.* Rio de Janeiro, 1949.

———. "O Banguê em Pernambuco no Século XIX," *Revista do Archivo Público* (Recife), Anos VII-X, nos. IX-XII (1952-1956), pp. 15-30.

———. *População e Açúcar no Nordeste do Brasil.* São Paulo, 1954.

Direction Générale de Statistique. *Annuaire Statistique du Brésil, 1ère année (1908-1912).* 2 volumes, Brésil, 1917.

Directoria Geral de Estatistica. *Idades da População Recenseada em 31 de Dezembro de 1890.* Rio de Janeiro, 1901.

———. *Indústria Assucareira. Usinas e Engenhos Centraes.* Rio de Janeiro, 1910.

———. "Recenseamento da população em 31 de Dezembro de 1900," *Relatório apresentado ao Dr. Miguel Calmon du Pin e Almeida, Ministro da Industria, Viação e Obras Públicas pelo Dr. José Luiz S. de Bulhões Carvalho.* Rio de Janeiro, 1908.

———. *Recenseamento do Brazil realizado em 1 de setembro de 1920.* 5 volumes, Rio de Janeiro, 1922-30.

———. *Relação dos proprietários dos Estabelecimentos Ruraes Recenseados no Estado de Pernambuco.* Rio de Janeiro, 1925.

———. *Relatório e Trabalhos Estatisticos apresentados ao Illm. e. Exm. Sr. Conselheiro Dr. Carlos Leoncio de Carvalho Ministro e Secretário de Estado dos Negócios do Imperio pelo Director Geral Conselheiro Manoel Francisco Correia em 20 de Novembro de 1878.* Rio de Janeiro, 1879.

———. *Sexo, raça e estado civil, nacionalidade, filiação, culto e analphabetismo da população recenseada em 31 de dezembro de 1890.* Rio de Janeiro, 1898.

Duncan, Julian Smith. *Public and Private Operation of Railways in Brazil.* New York, 1932.

Dunshee de Abranches. *Governos e Congressos da Republica, 1889-1917.* 2 volumes, Rio de Janeiro, 1918.

Eichner, Alfred S. *The Emergence of Oligopoly. Sugar Refining as a Case Study.* Baltimore and London, 1969.

Eisenberg, Peter L. "Abolishing Slavery. The Process on Pernambuco's Sugar Plantations," *Hispanic American Historical Review*, v. 52, no. 4 (November 1972).

————. "Falta de Imigrantes. Um Aspecto do Atraso Nordestino," *Revista de Historia* (São Paulo), January-March 1973.

Eisenstadt, S. M. *Modernization: Protest and Change.* Englewood Cliffs, N.J., 1966.

Elkins, Stanley. *Slavery. A Problem in American Institutional and Intellectual Life.* Chicago, 1959.

Ely, Roland. *Cuando Reinaba Su Majestad El Azúcar.* Buenos Aires, 1963.

Engels, Frederick. "The Origin of the Family, Private Property, and the State," in Karl Marx and Frederick Engels, *Selected Works.* 2 volumes, Moscow, 1958, v. II, pp. 170-327. First published 1884.

Engenhos Centrais no Brasil. Translated by Pereira Lima. Rio de Janeiro, 1877.

Estatutos do Syndicato Agricola Regional de Gamelleira, Amaragy, Bonito e Escada. Pernambuco, 1905.

Estatuto da União dos Syndicatos Agricolas de Pernambuco approvado em Assembléia Geral dos Syndicatos a 6 de março de 1906. Recife, 1906.

Facó, Rui. *Cangaceiros e Fanaticos.* 2nd edition, Rio de Janeiro, 1965.

Fairbanks, George Eduardo. *Observações sôbre o commércio de Assucar, e o Estado presente desta Industria.* Bahia, 1847.

Fairrie, Geoffrey. *Sugar.* Liverpool, 1925.

Faoro, Raymundo. *Os Donos do Poder. Formação do Patronato Político Brasileiro.* Porto Alegre, 1958.

Fernandes, Florestan. *Comunidade e Sociedade no Brasil.* São Paulo, 1972.

————. *The Negro in Brazilian Society.* New York, 1969. First published 1965.

Fernandes Gama, José Bernardo. *Memórias Históricas da Provincia de Pernambuco,* 2 volumes, Pernambuco, 1844.

260   THE SUGAR INDUSTRY IN PERNAMBUCO

Fernandes Lopes, João. *Colônias Industriaes Destinadas à Disciplinas, correcção e educação dos vagabundos regenerados pela hospitalidade e trabalho ou Exemplos fecundos das medidas preventivas contra a mendicidade e vagabundagem empregada na França, Suissa, Allemanha, Hollanda, Inglaterra e Estados Unidos por meio de regulamentos até 1889.* Pernambuco, 1890.

Ferreira Lima, Heitor. *História Político-Econômica e Industrial do Brasil.* São Paulo, 1970.

Ferreira Reis, Artur Cezar. "Inquietações do Norte," in Buarque de Holanda, *História Geral da Civilização Brasileira,* tomo I, v. 2, pp. 380-393.

Ferreira Soares, Sebastião. *Notas Estatísticas sôbre a Produção Agrícola e Carestia dos Gêneros Alimentícios no Império do Brasil.* Rio de Janeiro, 1860.

Ferrez, Gilberto. *Velhas Fotografias Pernambucanas, 1841-1900.* Recife, n.d.

Fialho, Anfrisio. *Don Pedro II, Empereur du Bresil. Notice Biographique.* Brussels, 1876.

———. *Impending Catastrophe. The C-S-F-of B-, ltd.* London, 1884.

———. *Um Terço de Século (1852-1885). Recordações.* Rio de Janeiro, 1885.

Figueira de Mello, Jerônymo Martiniano. *Ensaio sôbre a Statística Civil e Política da Provincia de Pernambuco.* Recife, 1852..

Fogel, Robert William, and Stanley L. Engerman. "The Economics of Slavery," in *idem., The Reinterpretation of American Economic History.* New York, 1971, pp. 311-341.

Foner, Philip S. *A History of Cuba and Its Relations with the United States.* 2 volumes, New York, 1962-63.

Fonseca Hermes Júnior, João Severiano da. *O Assucar como factor importante da riqueza pública do Brasil.* Rio de Janeiro, 1922.

Forman, Shepard. "Disunity and Discontent: A Study of Peasant Political Movements in Brazil," in Ronald H. Chilcote (editor), *Protest and Resistance in Angola and Brazil. Comparative Studies.* Berkeley and Los Angeles, 1972, pp. 183-205.

Freire Aquino Fonseca, Célia. "Rotas, Portos, Comércio e a Formação do Compléxo Açucareiro em Pernambuco," in *Anais do V Simpósio Nacional dos Professores Universitários de História. Portos, Rotas e Comercio.* 2 volumes, São Paulo, 1971. v. l, pp. 345-365.

Freitas, Octavio de. *O Clima e a Mortalidade da Cidade do Recife.* Recife, 1905.

Freyre, Gilberto. *Um Engenheiro Francês no Brasil.* Rio de Janeiro, 1940.

―――. *O Escravo nos anúncios de jornais brasileiros do Século XIX.* Recife, 1963.

―――. *Inglêses no Brasil. Aspectos da influência britânica sôbre a vida, a paisagem e a cultura do Brasil.* Rio de Janeiro, 1948.

―――. *The Mansions and the Shanties (Sobrados e Mucambos) The Making of Modern Brazil.* Translated by Harriet de Onís. New York, 1963. First published 1936.

―――. *The Masters and the Slaves (Casa Grande e Senzala). A Study in the Development of Brazilian Civilization.* Translated by Samuel Putnam. 2nd edition revised, New York, 1956. First published 1933.

―――. *Nordeste, Aspectos da Influência da Cana sôbre a Vida e a Paisagem do Nordeste do Brasil.* 4th edition, Rio de Janeiro, 1967. First published 1937.

―――. *Ordem e Progresso.* 2 volumes, Rio de Janeiro, 1959.

―――. *Order and Progress. Brazil from Monarchy to Republic.* Edited and translated by Rod W. Horton, New York, 1970.

―――. *Vida Social no Brasil nos Meados do Século XIX.* translated by Waldemar Valente. Recife, 1964. First published as "Social Life in Brazil in the Middle of the Nineteenth Century," *Hispanic American Historical Review*, v. V (1922), pp. 597-630.

Furtado, Celso. *Economic Development of Latin America. A Survey from Colonial Times to the Cuban Revolution.* Translated by Suzette Machado. Cambridge, England, 1970.

―――. *The Economic Growth of Brazil. A Survey from Colonial to Modern Times.* Translated by Ricardo W. Aguiar and Eric Drysdale, Berkeley and Los Angeles, 1963. First published as *Formação Econômica do Brasil,* 1959.

Galloway, J. H. "The Last Years of Slavery on the Sugar Plantations of Northeastern Brazil," *Hispanic American Historical Review*, v. 51, no. 4 (November 1971), pp. 586-605.

―――. "The Sugar Industry of Pernambuco During the Nineteenth Century," *Annals of the Association of American Geographers*, v. LVIII, no. 2 (June 1968), pp. 285-303.

Gardner, George. *Viagens no Brasil principalmente nas provincias do norte e nos distritos de ouro e do diamante durante os annos de 1836-1841.* Translated by Albertino Pinheiro. São Paulo, 1943. First published 1846.

Genovese, Eugene D. *The Political Economy of Slavery. Studies in the Economy and Society of the Slave South.* New York, 1967.

————. *The World the Slaveholders Made: Two Essays in Interpretation.* New York, 1969.

Góes, Raul de. *Um Sueco Emigra para o Nordeste.* 2nd edition, Rio de Janeiro, 1964.

Gomes de Mattos, Antonio. *Os Engenhos Centraes e o Sr. A. G. de Mattos. Colleção de artigos publicados no Jornal de Commércio de julho à agôsto de 1882.* Rio de Janeiro, 1882.

————. *Esbôço de um Manual para os Fazendeiros de Assucar no Brazil.* Rio de Janeiro, 1882.

Gonçalves e Silva. *O Assucar e o Algodão em Pernambuco.* Recife, 1929.

Gonsalves de Mello (Neto), J. A. "Trabalhadores Belgas em Pernambuco (1859-63)," *Boletim do Instituto Joaquim Nabuco de Pesquisas Sociais,* no. 8 (1959), pp. 13-37.

Gorringe (USN), Lieutenant Commander Henry Honeychurch. *The Coast of Brazil,* v. I, *From Cape Orange to Rio de Janeiro.* Washington, 1875.

Goulart, José Alípio. *Transportes nos Engenhos de Açúcar.* Rio de Janeiro, 1959.

Goulart, Mauricio. *Escravidão Africana no Brasil (Das origens à extinção do tráfico).* São Paulo, 1949.

Governador do Estado de Pernambuco, *Relatórios.* Various titles, also called *Falas* and *Mensagens.* Recife, 1890-1910.

Graham, Richard. "Brazilian Slavery Reconsidered." *Journal of Social History* (New Brunswick, New Jersey), v. 3, no. 4 (Summer 1970), pp. 431-453.

————. *Britain and the Onset of Modernization in Brazil, 1850-1914.* Cambridge, England, 1968.

————. "Landowners and the Overthrow of the Empire," *Luso-Brazilian Review,* (Madison, Wisconsin) v. VII, no. 2 (December 1970), pp. 44-56.

Great Britain. Parliament. *Parliamentary Papers.* House of Commons by Command. *Accounts and Papers.* London, 1844-1914.

Great Western of Brazil Railway Company, Limited. *Report of the Directors and Statement of Accounts . . . to be Submitted to the Shareholders at the Annual General Meeting.* London, 1886-1909.

Guerra, Flávio. "A arte de amealhar dinheiro," *Diário de Pernambuco,* May 7, 1967, Caderno V., pp. 10-11.

————. *História de Pernambuco.* 2 volumes, Rio de Janeiro, 1966.

————. *Idos do Velho Açúcar.* Recife, 1966.

Gunder Frank, Andre. *Capitalism and Underdevelopment in Latin America: Historical Studies of Chile and Brazil.* New York, 1967.

Hadfield, William. *Brazil, the River Plate, and the Falkland Islands.* London, 1854.

Hall, Gwendolyn Midlo. *Social Control in Slave Plantation Societies. A Comparison of St. Domingue and Cuba.* Baltimore, 1971.

Haring, C. H. *Empire in Brazil: A New World Experiment with Monarchy.* Cambridge, Massachusetts, 1958.

Henderson, James. *A History of the Brazil. Comprising its Geography, Commerce, Colonization, Aboriginal Inhabitants, etc., etc., etc.* London, 1821.

Hewitt, Cynthia N. "The Peasant Movement of Pernambuco, Brazil: 1961-64," in Henry A. Landsberger (editor), *Latin American Peasant Movements.* Ithaca, N.Y., 1969, pp. 374-398.

Hobsbawm, E. J. *Bandits.* Harmondsworth, England, 1972. First published 1969.

Holloway, Thomas. "Condições do mercado de trabalho e organização do trabalho nas plantações na economia caffeira de São Paulo, 1885-1915. Uma análise preliminar," *Estudos Econômicos* (São Paulo), v. 2, no. 6 (December 1972), pp. 145-177.

Ianni, Octavio. *Raças e Classes Sociais no Brasil.* Rio de Janeiro, 1966.

"Idéa da População da Capitania de Pernambuco, e das suas anexas, extensão de suas Costas, Rios e Povoações notaveis, Agricultura, numero dos Engenhos, contractos e Rendimentos Reaes, augmento que estes tem tido etc., desde o anno de 1774 em que tomou posse do Governo das mesmas Capitanias o Governador e Capitam Geral General Jozé Cesar de Menezes," *Annaes da Bibliotheca Nacional do Rio de Janeiro*, v. 40 (1918), pp. 1-111.

Iglesias, Francisco. "Vida Política, 1848-1868," in Buarque de Holanda, *História Geral da Civilização Brasileira*, tomo II, v. 3, pp. 9-112.

Imlah, Albert. *Economic Elements in the Pax Brittanica. Studies in British Foreign Trade in the Nineteenth Century.* Cambridge, Massachusetts, 1958.

Império do Brasil na Exposição Universal de 1873 em Vienna d'Austria. Rio de Janeiro, 1873.

Informações sôbre o Estado de Lavoura. Rio de Janeiro, 1874.

Instituto do Açúcar e do Àlcool. *Congressos Açucareiros no Brasil.* Rio de Janeiro, 1949.

Instituto Brasileiro de Geografia e Estatística, Departmento Estadual de Estatística, *Anuário Estatístico de Pernambuco*, ano IV (1930), ano XI (1942), ano XII (1946), ano XIX (1964). Recife.

Instituto Brasileiro da Geografia e Estatística, *Encyclopedia dos Municipios Brasileiros*. 36 volumes, Rio de Janeiro, 1957-58.

Jaguaribe, Helio. *Economic and Political Development. A Theoretical Approach and a Brazilian Case Study*. Cambridge, Massachusetts, 1968. First published as *Desenvolvimento Econômico e Desenvolvimento Político*, 1962.

James, Preston. *Latin America*. 3rd edition, New York, 1959. First published 1942.

Jenks, Leland. *Our Cuban Colony*. New York, 1928.

Kidder, D. P. *Sketches of Residence and Travels in Brazil*. 2 volumes, Philadelphia, 1845.

Klein, Herbert S. "The Internal Slave Trade in Nineteenth-Century Brazil: A Study of Slave Importations into Rio de Janeiro in 1852," *Hispanic American Historical Review*, v. 51, no. 4 (November 1971), pp. 567-585.

———. "Nineteenth-Century Brazil," in David W. Cohen and Jack P. Greene (editors), *Neither Slave nor Free: The Freedman of African Descent in the Slave Societies of the New World*. (Baltimore, 1972), pp. 309-334.

Knight, Franklin W. *Slave Society in Cuba During the Nineteenth Century*. Madison, Wisc., 1970.

Koster, Henry. *Travels in Brazil*. 2 volumes, Philadelphia, 1817.

Leff, Nathaniel. *The Brazilian Capital Goods Industry*. Cambridge, Mass., 1968.

———. "Desenvolvimento econômico e desigualdade regional; origens do caso brasileiro," *Revista brasileira de economia* (Rio de Janeiro), ano 26, no. 1 (January-March 1972), pp. 3-21.

———. "Economic Development and Regional Inequality: Origin of the Brazilian Case," *Quarterly Journal of Economics*, v. LXXXVI (May 1972), pp. 234-262.

———. "Economic Retardation in Nineteenth-Century Brazil," *Economic History Review*, Second Series, v. XXV, no. 3 (August 1972), pp. 489-507.

Lins do Rêgo, José. *Banguê*. Rio de Janeiro, 1934.

Lei e Regulamento da Reforma Hypothecaria estabelecendo as bases da Sociedades de Crédito Real. Rio de Janeiro, 1865.

Lima, Hermes. *Tobias Barreto (A epoca e o Homem)*. São Paulo, 1939.

Lira, Luis de. "As Bravuras de Antonio Silvino em honra de um velho amigo." n.p., n.d.

Manchester, Alan K. *British Preëminence in Brazil: Its Rise and Decline.* Chapel Hill, North Carolina, 1933.

Mansfield, Charles Blachford. *Paraguay, Brazil, and the River Plate, Letters Written in 1852-1853.* Cambridge, England, 1856.

Marc, Alfred. *Le Brésil. Excursion à Travers des 20 Provinces.* 2 volumes, Paris, 1890.

Marx, Karl. *Capital.* Translated by Samuel Moore and Edward Aveling, and edited by Frederick Engels. New York, 1906. First published 1867.

Mathieson, William Law. *Great Britain and the Slave Trade, 1839-1865.* London, 1929.

Medeiros, Coriolano de. "O Movimento da Abolição no Nordeste," *Livro do Nordeste Commemorativo do Primeiro Centenário do Diário de Pernambuco 1825-1925.* Recife, 1925, pp. 92-96.

Medeiros de Sant'Ana, Moacyr. *Contribuição à História do Açúcar em Alagoas.* Recife, 1970.

Melo, Mário. *Pau d'Alho: Geographia Physica e Política.* Recife, 1918.

――――. *Síntese Cronológica de Pernambuco.* Recife, 1942.

Melo Meneses, Diogo de, and Gilberto Freyre (editors). *O Velho Felix e suas "Memórias de um Cavalcanti."* Rio de Janeiro, 1959.

Milet, Henrique Augusto. *Auxilio à Lavoura e Crédito Real.* Recife, 1876.

――――. *A Lavoura da Canna de Assucar,* Pernambuco, 1881.

――――. *Os Quebra Kilos e a Crise da Lavoura.* Recife, 1876.

Ministerio da Agricultura, Industria e Commercio, Serviços de Inspecção e Defesa Agricolas. *Questionários sôbre as condições da agricultura dos 173 municipios do Estado de São Paulo. De abril de 1910 a janeiro de 1912.* Rio de Janeiro, 1913.

Ministro de Estado dos Negócios da Indústria, Viação e Obras Públicas. *Relatórios apresentados ao Presidente da República dos Estados Unidos do Brazil,* Rio de Janeiro, 1893-96.

Ministro e Secretario do Estado dos Negócios da Agricultura, Commércio e Obras Públicas. *Relatórios apresentados à Assembléa Geral.* Rio de Janeiro, 1861-92.

Mintz, Sidney W. "Labor and Sugar in Puerto Rico and Jamaica, 1800-1850," *Comparative Studies in Society and History,* v. I, no. 3 (March 1959), pp. 273-280, reprinted in Laura Foner and Eugene D. Genovese (editors), *Slavery in the New World: A Reader in Comparative History.* Englewood Cliffs, N.J., 1969, pp. 170-177.

———. "Review of Stanley M. Elkins' Slavery," *American Anthropologist*, v. LXIII (June 1961), pp. 579-587, reprinted as "Slavery and Emergent Capitalism," in Foner and Genovese, *Slavery in the New World*, pp. 27-37.

Moraes, Evaristo de. *A campanha abolicionista, 1879-1888*. Rio de Janeiro, 1924.

Moreno Fraginals, Manuel. *El Ingenio: El Complejo Económico Social Cubano del Azúcar*, Tomo 1 (1760-1860). La Habana, 1964.

Mota, Leandro. *Violeiros do Norte*. 3rd edition, Fortaleza, 1962.

Nabuco, Joaquim. *O Abolicionismo. Conferências e Discursos Abolicionistas*. São Paulo, 1949. First published 1883.

———. *Minha Formação*. Rio de Janeiro, 1957. First published 1900.

Netto Campello. *História Parlamentar de Pernambuco*. Recife, 1923.

North, Douglass C. *Growth and Welfare in the United States*. Englewood Cliffs, N.J., 1966.

Noticia acerca da Industria Assucareira no Brazil. Rio de Janeiro, 1877.

Novais, Fernando A. "O Brasil nos Quadros do Antigo Sistema Colonial," in Carlos Guilherme Mota, *Brasil em Perspectiva*. São Paulo, 1968. pp. 47-63.

Noya Pinto, Virgilio. "Balanço das Transformações Econômicas no Século XIX," in Mota, *Brasil em Perspectiva*, pp. 126-145.

Oliveira Lima. *Memórias (Estas minhas reminiscências . . .)*. Rio de Janeiro, 1937.

Oliveira Torres, João Camillo de. *Os Construtores do Império. Ideais e lutas do Partido Conservador Brasileiro*. São Paulo, 1968.

———. *A democracia coroada: teoria política do império do Brasil*. Rio de Janeiro, 1957.

Oliveira Vianna, F. J. "Resumo Histórico dos Inquéritos Censitários Realizados no Brazil," Directoria Geral da Estatística, *Recenseamento do Brazil realizado em 1 de setembro de 1920*, v. I, pp. 404-483.

Ónody, Oliver. *A Inflação Brasileira (1820-1958)*. Rio de Janeiro, 1960.

Osório de Cerqueira, José. *La Province de Pernambuco*. Pernambuco, 1888.

Paasche, Hermann. *Zuckerindustrie und Zuckerhandel der Welt*. Jena, 1891.

Padua, Ciro T. de. "Um Capítulo da História Econômica do Brasil," *Revista do Arquivo Municipal* (São Paulo), Ano XI, v. C. (January-February 1945), pp. 135-190.

Paes de Andrade, Luiz de Carvalho. *Questões Econômicas em Relação à Província de Pernambauco*. Recife, 1864.

Paes Barreto, Carlos Xavier. "A estirpe dos Lins," *Revista do Instiuto Arqueológico Histórico e Geográfico Pernambucano*. v. XLVI (1961) pp. 209-215.

Parry, J. H. and P. M. Sherlock. *A Short History of the West Indies*. 3rd edition, London, 1971. First published 1956.

Pereira da Costa, F. A. *Anais Pernambucanos*. 10 volumes, Recife, 1949-66.

————. "Noticia sôbre as instituições de crédito bancário em Pernambuco, offerecida à Benemerita Associação Commercial e Beneficente," in *Relatório da Direcção da Associação Commercial Beneficente de Pernambuco apresentado à Assembléa Geral da mesma em 10 de agôsto de 1898*, pp. 95-112.

————. "Origens Históricas da Industria Assucareira em Pernambuco," in *Trabalhos da Conferência Assucareira*, pp. I-XLIX.

————. "Origens historicas do Municipio da Escada," *A Escada*, April 7, 1904.

Peres, Apollonio, and Manuel Machado Cavalcanti. *Industrias de Pernambuco*, Recife, 1935.

Peres, Gaspar, and Apollonio Peres. *A Industria Assucareira em Pernambuco*. Recife, 1915.

Pernambuco. Biblioteca Pública. *Documentação Histórica Pernambucana. Sesmarias*. 3 volumes, Recife, 1959.

Picanço, Francisco. *Diccionario de Estradas de Ferro*. 2 volumes, Rio de Janeiro, 1891-92.

Pinto, Adelia. *Um livro sem título. (Memórias de uma Provinciana)*. Rio de Janeiro, 1962.

Pinto, Estevão. *A Associação Comercial de Pernambuco. Livro Comemorativo de seu primeiro centenário (1839-1939)*. Recife, 1940.

————. *História de Uma Estrada-de-Ferro do Nordeste*. Rio de Janeiro, 1949.

Pires Ferreira, Júlio. *Almanack de Pernambulco para o anno de 1899, 1900, 1901*, Recife, 1899-1901.

Poppino, Rollie. *Brazil, The Land and the People*. New York, 1968.

Porter, Robert P. *Industrial Cuba*. New York, 1899.

Prado, Eduardo. *A Illusão Americana*. 3rd edition, São Paulo, 1961. First published 1893.

Prado Júnior, Caio, *Formação do Brasil Contemporâneo. Colônia*. 5th edition, São Paulo, 1957. First published 1942.

————. *História Econômica do Brasil*. 11th edition, São Paulo, 1969. First published 1945.

Presidente da Provincia de Pernambuco. *Relatórios*. Various titles, also called *Fallas* and *Mensagens*. Recife, 1844-89.

Prinsen Geerligs, H. C. *The World's Cane Sugar Industry. Past Present*. Manchester, 1912.

Projecto de Receita Provincial organizado por ordem do Excm. Desembargador José Manoel de Freitas, Digníssimo Presidente desta Província, pelo Administrador do Consulado Provincial Bacharel Francisco Amynthas de Carvalho Moura. Pernambuco, 1884.

Proposta e Relatório apresentados à Assembléa Geral Legislativa pelo Ministro e Secretário de Estado dos Negócios Estrangeiros e Interino da Fazenda Barão de Cotegipe. Rio de Janeiro, 1877.

Quintas, Amaro. "A agitação republicana no Nordeste," in Buarque de Holanda, *História Geral da Civilização Brasileira*, tomo II, v. 1, pp. 207-237.

————. "O Nordeste, 1825-1850," in Buarque de Holanda, *História Geral da Civilização Brasileira*, tomo II, v. 2, pp. 193-311.

————. *O Sentido Social da Revolução Praieira*, Rio de Janeiro, 1967.

Raffard, Henri. *O Centro da Indústria e Commércio de Assucar no Rio de Janeiro*. Rio de Janeiro, 1892.

Raiz, Jovino da. "O trabalhador negro no tempo do banguê comparado com o trabalhador negro no tempo das uzinas de assucar," *Estudos Afro-Brasileiros. Trabalhos apresentados ao 1º Congresso Afro-Brasileiro, reunido no Recife em 1934*. Rio de Janeiro, 1935, pp. 191-194.

Recenseamento da População do Império do Brasil a que se procedeu no Dia 1 de Agôsto de 1872. Quadros Estatísticos. 23 volumes, Rio de Janeiro, 1873-76.

Recife and San Francisco Railway Company (Limited). *Report of the Proceedings at the . . . Ordinary General Meeting of the Shareholders*. London, 1860-94.

"Regime dos Ventos," *Porto do Recife*, Anno 1, no. 1, (August 1933).

Relatório apresentado ao Exmo. Sr. Antonio Gonçalves Ferreira, Governador do Estado, pelo Secretário da Indústria, Obras Públicas, Agricultura, Commercio e Hygiene. Recife, 1902.

Relatório da Commissão Central de Socorros aos indigentes victimas da secca. Pernambuco, 1878.

Relatório da Commissão dirigida por Joaquim d'Aquino Fonseca, apresentado ao Excellentissimo Sr. Conselheiro Dr. José Bento da Cunha Figueiredo no 10 de janeiro de 1856. Pernambuco, 1856.

Relatório do Inspector da Fazenda Provincial de Pernambuco José Pedro da Silva, apresentado no 3⁰ de Fevereiro de 1852 ao Excellentíssimo Sr. Victor de Olivera, presidente da mesma provincia. Pernambuco, 1852.

Ribeiro Lamego, Alberto. *O Homem e o Brejo.* Rio de Janeiro, 1945.

Ribeiro Pessôa Junior, Cyro Dioclesiano. *Estudo Descriptivo das Estrades de Ferro do Brazil Precedido da Respectiva Legislaçao.* Rio de Janeiro, 1886.

Robock, Stefan H. *Brazil's Developing Northeast: A Study of Regional Planning and Foreign Aid.* Washington, 1963.

Roche, Jean. *La colonisation allemande et le Rio Grande do Sul.* Paris, 1959.

Rodrigues, José Honório. *Índice anotado da Revista do Instituto Arqueológico Histórico e Geográfico Pernambucano.* Recife, 1961.

———. "A Literatura Brasileira sôbre Açúcar no Século XIX," *Brasil Açucareiro,* v. XIX, no. 5 (May 2, 1942), pp. 16-38.

———. "A Revolução Industrial Açucareira. Os Engenhos Centrais," *Brasil Açucareiro,* v. XXVII, nos. 2-4 (February-April 1946), pp. 179-182, 229-233, 392-397.

Roett, Riordan. *The Politics of Foreign Aid in the Brazilian Northeast.* Nashville, 1972.

Rosa e Silva Neto, J. M. *Contribuição ao Estudo da Zona da Mata em Pernambuco.* Recife, 1966.

Russell-Wood, A. J. R. "Colonial Brazil," in Cohen and Greene, *Neither Slave nor Free,* pp. 84-133.

Sagra, Ramón de la. *Cuba 1860. Selección de artículos sôbre agriculture cubana.* La Habana, 1963.

Salles, Alinio de. "O Verdadeiro Responsável pela Introdução da 'Cana Caiana' em nosso País," *Brasil Açucareiro,* v. LXXIII, no. 4 (October 1968), pp. 46-48.

Santos Dias Filho, Manoel Antônio dos. "Industria Assucareira," *Boletim do Ministério da Agricultura, Indústria, e Commercio,* Anno I, no. 5 (1913), pp. 59-63.

Schorer Petrone, Maria Teresa. "Imigração assalariada," in Buarque de Holanda, *História Geral da Civilização Brasileira,* tomo II, v. 3, pp. 274-296.

———. *A lavoura canavieira em São Paulo.* São Paulo, 1968.

Sette, Mário. *Arruar. História Pitoresca do Recife Antigo.* 2nd edition, Rio de Janeiro, n.d. First published 1948.

Simonsen, Roberto. "As consequências econômicas de abolição," *Jornal do Comercio* (São Paulo) May 8, 1938, reprinted in *Revista do ar-*

quivo municipal (São Paulo), ano iv, v. XLVII (May 1938), pp. 257-268.

―――. *História Econômica do Brasil, 1500-1820.* 4th edition, São Paulo, 1962. First published 1937.

Singer, Paul. *Desenvolvimento Econômico e Evolução Urbana.* São Paulo, 1968.

Sitterson, J. Carlyle. *Sugar Country: The Cane Industry in the South, 1753-1950.* Lexington, Kentucky, 1953.

Smith, Herbert H. *Brazil: The Amazons and the Coast.* New York, 1879.

Smith, T. Lynn (editor). *Agrarian Reform in Latin America.* New York, 1965.

Sociedade Auxiliadora da Agricultura de Pernambuco. *Acta da Sessão da Assembléa Geral de 23 de abril de 1877 e Relatório Annual do Gerente o Senhor Ignácio de Barros Barreto.* Recife, 1877.

―――. *Acta da Sessão Solemne da Assembléa Geral de 28 de setembro de 1882 e Relatório Annual do Gerente Ignácio de Barros Barreto.* Recife, 1882.

―――. *Estatutos para o Banco Auxiliador d'Agricultura, Instituição de Credito Real e Agricola.* Recife, 1883.

―――. *Relatório Annual apresentado na sessão de 4 de julho de 1878 da Assembléa Geral pelo Gerente Ignácio de Barros Barreto e a Acta da mesma sessão.* Recife, 1878.

―――. *Trabalhos do Congresso Agrícola do Recife em outubro de 1878.* Recife, 1879.

Sociedade Nacional da Agricultura. *Legislação Agrícola do Brasil, Primeiro Período. Império (1808-1889).* Rio de Janeiro, 1910.

Souto Maior, Mario. *Antônio Silvino, Capitão do Trabuco.* Rio de Janerio, 1971.

Souza, Amaury de. "The *Cangaço* and the Politics of Violence in Northeast Brazil," in Chilcote, *Protest and Resistance in Angola and Brazil,* pp. 109-131.

Souza, Eusébio de. "O Quadro Histórico," in Raimundo Girao, *A abolição no Ceará.* Fortaleza, 1956.

Souza Gomes, Luíz. *Dicionário Econômico e Financeiro.* 8th edition, Rio de Janeiro, 1966.

Souza Leão Pinto, Eudes de. *Cana-de-açúcar.* Rio de Janeiro, 1965.

Souza e Silva, Joaquim Norberto. "Investigações sôbre os Recenseamentos de População Geral do Império e de Cada Província e per si tentados desde os Tempos Coloniais até hoje," in *Relatório apresentado à Assembléa Geral na segunda sessão da décima quarta legislatura pelo Ministro e Secretário de Estado dos Negócios do*

Império, Paulino José Soares de Souza, Appendix D. Rio de Janerio, 1870, reprinted by the Serviço Nacional de Recenseamento, *Documentos Censitários,* Série B, no. 1. Rio de Janeiro, 1951.

Stampp, Kenneth M. *The Era of Reconstruction, 1865-1877.* New York, 1965.

Stein, Stanley J. *The Brazilian Cotton Manufacture. Textile Enterprise in an Underdeveloped Area, 1850-1950.* Cambridge, Massachusetts, 1957.

————. *Vassouras. A Brazilian Coffee County, 1850-1900.* Cambridge, Massachusetts, 1957.

————. and Barbara H. Stein. *The Colonial Heritage of Latin America. Essays on Economic Dependence in Perspective.* New York, 1970.

Stonier, Alfred W. and Douglas C. Hague. *A Textbook of Economic Theory.* New York, 1961.

Suzannet, Conde de. *O Brasil em 1845. (Semelhanças e Differenças após um século).* Translated by Marcia de Moura Castro. Rio de Janeiro, 1957.

Tamer, Alberto. *O Mesmo Nordeste.* São Paulo, 1968.

Tannenbaum, Frank. *Slave and Citizen, the Negro in the America.* New York, 1946.

Taunay, Affonso de Escragnolle. *História do Café no Brasil.* 15 volumes, Rio de Janeiro, 1939-43.

————. *Pequena História do Café no Brasil (1727-1937).* Rio de Janeiro, 1945.

Thomas, Hugh. *Cuba: The Pursuit of Freedom.* New York, 1971.

Tischendorf, Alfred. "The Recife and San Francisco Railway Company, 1854-1860," *Inter-American Economic Affairs* (Washington, D.C.), v. XIII, no. 4 (Spring 1960), pp. 87-94.

Tollenare, L. F. *Notas Dominicais tomadas durante uma residencia em Portugal e no Brazil nos annos de 1816, 1817 e 1818. Parte relativa a Pernambuco.* Translated by Alfredo de Carvalho, Recife, 1905.

Toplin, Robert Brent. *The Abolition of Slavery in Brazil.* New York, 1972.

————. "Upheaval, Violence, and the Abolition of Slavery in Brazil: The Case of São Paulo," *Hispanic American Historical Review,* v. XLIX, no. 4 (November 1969) pp. 639-655.

Trabalhos da Conferência Assucareira do Recife (2º do Brasil). Recife, 1905.

U.S. Congress, House of Representatives, *Executive Documents.* Washington, 1850-89.

————. *Reports for the Year. . . on the Trade of the Consular District of Pernambuco*. Washington, 1857-96.

U.S. Department of State. *Reports from the Consuls of the United States on the Commerce, Manufactures, etc. of Their Consular Districts*. Washington, 1880-1910.

Vasconcellos, José de. *Almanack Administrativo, Mercantil e Industrial da Provincia de Pernambuco para o Anno de 1860, 1861, 1862*. Recife, 1860-62.

Vasconcellos Galvão, Sebastião. *Diccionario Chorographico, Historico e Estatistico de Pernambuco*. 4 volumes, Rio de Janeiro, 1908-27. Various editions.

Vasconcelos Sobrinho. *As Regiões Naturais de Pernambuco, O Meio e a Civilização*, Rio de Janeiro, 1949.

Vasconcellos Torres. *Condições de vida do trabalhador no agro-indústria do açucar*. Rio de Janeiro, 1945.

Veiga, Luiz Francisco da. *Livro do Estado Servil e Respectiva Libertação contendo a Lei de 28 de Setembro de 1871 e os decretos e avisos expedidos pelos Ministros da Agricultura, Fazenda, Justiça, Império e Guerra desde aquella data até 31 de Dezembro de 1875*. Rio de Janeiro, 1876.

Veiga Cabral, Mario da. *Coreografia do Brasil*. 28th edition, Rio de Janeiro, 1947.

Verissimo de Toledo. *Almanack Administrativo Mercantil, Agrícola, e Industrial do Estado de Pernambuco para 1893, 1894, 1895*. Recife, 1893-95.

Vilela Luz, Nícia. *A Luta pela Industrialização do Brazil (1808 à 1930)*. São Paulo, 1961.

————. "O papel das classes médias brasileiras no movimento republicano," *Revista de História* (São Paulo), ano XV, v. XXVIII (1964), pp. 13-27.

Viotti da Costa, Emilia. *Da Senzala à Colônia*. São Paulo, 1966.

————. "O escravo na grande lavoura," in Buarque de Holanda, *História Geral da Civilização Brasileira*, tomo II, v. 3, pp. 135-188.

Viveiros, Jeronimo de. "O Açúcar através do periódico 'O Auxiliador da Industria Nacional'," *Brasil Açucareiro*, v. XXVII, no. 4 (April 1946), pp. 116-118.

Walle, Paul. *Au Brésil du Rio São Francisco à l'Amazone*. Paris, 1912.

Waterton esq., Charles. *Wanderings in South America, the Northwest of the United States, and the Antilles, in the Years 1812, 1816, 1820, and 1824*. 2nd edition, London, 1828.

Wanderley de Araújo Pinho, José. *História de um Engenho do Recôncavo, Matoim-Novo Caboto Freguezia, 1522-1944*. Rio de Janeiro, 1946.

Watts, Alfredo J. "A Colônia Inglêsa em Pernambuco," *Revista do Instituto Arqueológico, Histórico e Geográfico Pernambucano*, v. XXXIX (1944), pp. 163-170.

Wells, James W. *Exploring and Travelling Three Thousand Miles Through Brazil From Rio de Janeiro to Maranhão*. 2 volumes, London, 1886.

Werneck Sodré, Nelson. *História da Burguesia Brasileira*. 2nd edition, Rio de Janeiro, 1967.

—————. *As Razões da Independência*. 2nd edition, Rio de Janeiro, 1969. First published 1965.

Wiederspahn, Henrique Oscar. "Dos Lins de Ulm e Augsburgo aos Lins de Pernambuco," *Revista do Instituto Arqueológico Histórico e Geográfico Pernambucano*, v. XLVI (1961) pp. 7-98.

"The World's Sugar Production and Consumption," in United States Treasury Department, Bureau of Statistics, *Summary of Commerce and Finance for November 1902*. Washington, D.C.

INDEX

Aberdeen Bill, 152
Abolition. *See* Slavery, abolition of
Afogados, 126
Africa, African. *See* Slavery, African importation of
Agricultural Commercial Association (ACA): expectations for railroads, 56; campaign against sugar export tax, 56; failure to establish bank, 77-78; demand for government loans to banks, 80; on central mills, 86; formation of, 141; on abolition, 170
Agricultural Congress. *See* Recife Agricultural Congress, 1878; Second Agricultural Congress
Agricultural syndicate movement, 143
Agriculture, cane: generally, 34-36; backwardness of, 32, 43; implements for, 34, 35; slave labor in, 34, 35, 36, 43; labor-intensive technology in, 34, 41-42; railroads used in, 35; deforestation in, 35, 36; fertility of soil in, 36; land use in, 36, 41, 49 (*see also* Land use); fertilization of soil in, 36, 39; soil destruction in, 36, 43; routinism in, 42-43; experimental farm, 49; reforestation, 114n43
Agriculture minister, 89-90
Agriculture ministry, 104-105n36
Água Preta: central mills in, 88, 93; humid mata region in, 125; railroad in, 125; slave population in, 12, 218; support of Colligação in, 27; sugar exports in, 31; railroad in, 52; humid mata region bounded by, 123; Agriculture minister from, 143; immigrant colony near, 200
Alagoas, 12, 31, 52, 80, 123, 143
Albuquerque Mello, José Maria de, 169, 170

Alcohol: and sugar exports compared, 28, 39; domestic demands for, 29; possible uses of, 29; lack of demand on world market for, 29. *See also* Rum
Alfredo, João, 80
Aliança, 123
Alves Branco Tariff, 47
Alves da Silva family, 132
Amarají: central mill in, 103; humid mata region in, 125; railroad in, 125
Amazonas, 12
Amerindians, 7
Amorim, Manoel João d', 78
Antilles, 204
Antônio Olinto, 53
Araújo family, 133
Argentina: sugar industry in, 24; beef imports from, 56
Arruda Beltrão, Antônio Carlos de, 103
Arruda Falcão, Francisco Dias de, 134
Arruda Falcão, João Francisco de, 134
Associação Acadêmica Promotora da Remissão dos Captivos, 162
Associação Comercial Beneficente de Pernambuco (ACBP): on sugar prices, 19; on reciprocal trade agreement with U.S., 22; on crude sugar export, 26; protest of harbor disadvantages, 59; campaign against sugar export tax, 66; criticism of government bonds, 70; demand for government loans to banks, 80; criticism of government bonds for usinas, 115n45; formation of, 141; meeting hall, 170; quoted, on Polish immigration, 199; criticism of immigration, 213
Associação de Socorros Mutuos e Lenta Emancipação dos Captivos, 162